Transboundary Water Resources in the Balkans

T0137988

NATO Science Series

A Series presenting the results of activities sponsored by the NATO Science Committee. The Series is published by IOS Press and Kluwer Academic Publishers, in conjunction with the NATO Scientific Affairs Division.

A. Life Sciences	IOS Press
B. Physics	Kluwer Academic Publishers
C. Mathematical and Physical Sciences	Kluwer Academic Publishers
D. Behavioural and Social Sciences	Kluwer Academic Publishers
E. Applied Sciences	Kluwer Academic Publishers
F. Computer and Systems Sciences	IOS Press
1. Disarmament Technologies	Kluwer Academic Publishers
2. Environmental Security	Kluwer Academic Publishers
3. High Technology	Kluwer Academic Publishers
4. Science and Technology Policy	IOS Press
5. Computer Networking	IOS Press

NATO-PCO-DATA BASE

The NATO Science Series continues the series of books published formerly in the NATO ASI Series. An electronic index to the NATO ASI Series provides full bibliographical references (with keywords and/or abstracts) to more than 50000 contributions from international scientists published in all sections of the NATO ASI Series.
Access to the NATO-PCO-DATA BASE is possible via CD-ROM "NATO-PCO-DATA BASE" with user-friendly retrieval software in English, French and German (WTV GmbH and DATAWARE Technologies Inc. 1989).

The CD-ROM of the NATO ASI Series can be ordered from: PCO, Overijse, Belgium

Transboundary Water Resources in the Balkans

Initiating a Sustainable Co-operative Network

edited by

Jacques Ganoulis
Aristotle University of Thessaloniki,
Thessaloniki, Greece

Irene Lyons Murphy
Colorado State University,
Fort Collins,
Colorado, U.S.A.

and

Mitja Brilly
University of Ljubljana,
Ljubljana, Slovenia

Kluwer Academic Publishers

Dordrecht / Boston / London

Published in cooperation with NATO Scientific Affairs Division

Proceedings of the NATO Advanced Research Workshop on
Transboundary Water Resources in the Balkans: Initiating Regional Monitoring Networks
Thessaloniki, Greece
11–15 October 1999

A C.I.P. Catalogue record for this book is available from the Library of Congress.

ISBN 0-7923-6556-9 (HB)
ISBN 0-7923-6557-7 (PB)

Published by Kluwer Academic Publishers,
P.O. Box 17, 3300 AA Dordrecht, The Netherlands.

Sold and distributed in North, Central and South America
by Kluwer Academic Publishers,
101 Philip Drive, Norwell, MA 02061, U.S.A.

In all other countries, sold and distributed
by Kluwer Academic Publishers,
P.O. Box 322, 3300 AH Dordrecht, The Netherlands.

Printed on acid-free paper

TABLE OF CONTENTS

Part III. The Status of Water Resources Monitoring in the Balkans

Part IV. Support for a Balkan Water Network

PREFACE

The 50 water resources specialists from all ten Balkan countries, and also France, the United Kingdom and the United States, who gathered in Thessaloniki in the autumn of 1999 for the Advanced Research Workshop (ARW) on transboundary network monitoring, worked with energy and concentration to achieve its goals. Funded by the NATO Scientific Affairs Division and organised by the Laboratory of Hydraulics and Environmental Engineering, Department of Civil Engineering, Aristotle University of Thessaloniki, the Workshop provided the opportunity to review common Balkan transboundary water problems (figure 1), discuss ways to resolve them, and determine how to support networks that would help restore regional water resources and protect them in the future.

Participants included scientists, government officials and technical experts, who gave generously of their time, responded to pre-Workshop requests for information, used state-of-the-art graphics for their presentations, discussed the need for the continuous exchange of information about shared rivers, lakes, and seas, and suggested ways in which this might be achieved.

These discussions have resulted in the establishment of the International Network of Water-Environment Centres for the Balkans (INWEB). Information on this significant achievement may be found on the INWEB website **http://socrates.civil.auth.gr/inweb**, which has been developed by the host institution, the Aristotle University of Thessaloniki, Greece. The website aims to promote and follow-up communication between interested parties, and to report on further activities of INWEB.

On behalf of the University and the Workshop organising committee I would like to thank all the participants for their dedication and continued support, as well as those who were unable to attend but still sent in articles of value and interest. We are grateful for the well-structured presentations of case studies on transboundary water monitoring, and to the working groups for rivers, lakes and coastal waters, which met separately to draft conclusions and recommendations. Thanks are also due to the visiting specialists from France, the UK, and the US, who made suggestions for the paths that networking should take in the future.

We are very grateful to the NATO Scientific Affairs Division for its financial support, and to the Research Committee of the Aristotle University of Thessaloniki for additional funding.

Generous and invaluable help came at all stages from the members of the organising committee. Especial thanks are due to Professor Evan Vlachos of Colorado State University, who initiated the idea of this ARW, and to Dr. Irene Lyons Murphy for her hard work during the preparation of the proposal, the Workshop itself, and for advice and invaluable assistance during the preparation of this book. Our appreciation also goes to Professor Mitja Brilly, University of Ljubljana, who served well as Co-Director of the ARW and Co-Editor of the book.

Thanks are also extended to staff and graduate students from the Aristotle University of Thessaloniki for all their help, and particularly to Katie Quartano for her professional and efficient handling of local arrangements at the hotel Philipion during the workshop, and her meticulous editorial assistance in preparing the final manuscript.

The greatest achievement of the Workshop is the fact that all participants agreed to promote, through state-of-the-art information and communication technologies, the establishment of INWEB. One of the main tasks of this network will be to create and maintain a database for water quantity and quality monitoring in the Balkans, complementary to the European Environment Agency's EUROWATERNET. The participants at the workshop agreed to establish a regional follow-up committee to launch INWEB and to assist in its future operation.

We are grateful for substantial funding from the Greek Ministry of the Environment to help us achieve the overall goals of INWEB, promote the concept of sustainable development and the protection of natural water resources, as well as enhance the development of a "culture of peace" in the Balkan region.

The continuation of the activities of this Workshop through the creation of INWEB is certainly a very promising and exciting prospect for sustainable water resources management in the Balkans.

J. Ganoulis
Thessaloniki, April 2000

Figure 1: Main rivers and country borders in the Balkans.

Part I: Introduction and Key Issues

WORKSHOP GOALS: A BALKAN MONITORING NETWORK

J. GANOULIS
Laboratory of Hydraulics
Department of Civil Engineering
Aristotle University of Thessaloniki
54006 Thessaloniki, Greece

I. L. MURPHY
Consultant
2005 37ᵗʰ St NW
Washington, DC, USA

1. Seeking a Balkan Agreement

The five countries previously part of the Yugoslav Republic and now in transition, together with Albania, Bulgaria, and Romania, all share acute transboundary water resource problems with each other and with the regional NATO countries of Greece and Turkey. Participants from each of these countries addressed these problems at the NATO Advanced Research Workshop held in Thessaloniki, Greece, in October 1999. The articles in this book are the fruits of their presentations and the vigorous discussions that followed them. They focus on the need to improve regional water resource management, by replacing inadequate and uncertain monitoring of their shared lakes, rivers, and lengthy coastlines. They recommend the monitoring of water availability, including its quality, and using state-of-the-art scientific principles. This is key to adequate and cost-effective rehabilitation and protection of essential resources.

Participants from the Balkan countries provided case histories of current monitoring at five transboundary sites. They were followed by experts, who reviewed the problems and policies of their respective countries with regard to the management of water resources monitoring, exploring problems and making recommendations for national and international action. Specialists from the European Environmental Agency, Colorado State University, and French research centres reviewed the general question of transboundary water monitoring and the possibilities for regional networks, as well as the negotiation techniques they require. The concluding articles emphasise the need for a Balkan international centre for developing and maintaining co-ordinated solutions to Balkan water management issues.

J. Ganoulis et al. (eds.), Transboundary Water Resources in the Balkans, 1–5.
© 2000 *Kluwer Academic Publishers.*

2. An Overview: Introducing Workshop Concepts

Two articles in Part I provide background for the case studies and reviews of current Balkan transboundary monitoring practices described in Parts II and III. Ganoulis describes the complexities of transboundary water resources management and suggests possible strategies for regional negotiations to resolve differences between countries regarding their water development and use goals. He analyses the technical and institutional approaches that could lead to agreements on the sharing of international waters. He suggests that effective implementation of such treaties should use a bottom-up approach, based on "regional partnerships". He further suggests that the creation of an international centre can provide a modus operandi for the development and maintenance of such partnerships.

The Brilly-Murphy article elaborates the need for improved development and sharing of both water quantity and quality data in transboundary waters throughout the Balkans. The authors point out that most Balkan countries increasingly are part of, or have access to, established electronic networks, which can improve transboundary water resource management. To date these networks have not been able to help resolve transboundary issues or to ensure improved water quality at the national or regional level, primarily because of the lack of comparable trend data.

The adoption of EEA standards and development of linkages with the Danube Programme's data monitoring network by all Balkan countries will improve opportunities for management of shared water resources. Long-term planning can produce cost-effective treatment and development policies in transboundary rivers, lakes, and streams. Firm commitments on the part of research institutes and government ministries can accomplish a great deal towards building this type of support. A regional centre should be adopted to encourage and co-ordinate their networking activities.

3. Balkan Transboundary Monitoring: Case Studies

The Part II series of case studies illuminates the potential for electronic monitoring networks in the Balkans and their contribution to ongoing co-operative water resource activities in the region. These articles include a range of countries and types of water resources and transboundary projects in five areas of the Balkans, from the Lake Ohrid conservation project in Albania and the Former Yugoslav Republic of Macedonia (FYROM) to the coastal regions of Turkey. Each explores the current need for and status of transboundary management of significant shared water resources.

The Sava River, a tributary of the Danube River, once the largest national river basin in the former Yugoslavia with a population of seven million is now international, its catchment area providing drainage for four countries: Slovenia, Croatia, Bosnia and Herzegovina, and Yugoslavia. Some national co-ordinated water management systems and construction projects are now divided between the independent states but the need for improvement is very great. A pilot study of the Sava River basin is proposed to provide for transboundary monitoring and other co-operative activities.

The Nestos River project, described by five members of the Department of Civil Engineering of the Aristotle University, Thessaloniki, Greece, seeks to resolve serious transboundary water management issues on the river, which it shares with Bulgaria. For the first time the two countries have carried out a co-ordinated programme to improve the protection of this economically vital area, which is in danger from pollution and large-scale hydraulic works. A thorough transboundary environmental study has been jointly undertaken by the Aristotle University of Thessaloniki in co-operation with the University of Sofia.

The Lake Ohrid Conservation Project is sponsored by the World Bank with the co-operation of the other Global Environment Facility (GEF) members, and the UN Development and Environment Programmes, and provides support for transboundary co-operative activities by Albania and FYROM. The authors of the article are participants in the project and affiliated with the Hydrobiological Institute of FYROM and the Albanian Hydrometeorological Institute. Lake Ohrid, a unique aquatic ecosystem known to scientists throughout the world for its rich biodiversity, has deteriorated in quality, and its ecosystem balance has been seriously disturbed in the last 50 years. Although some protective measures have been undertaken, the present state of the lake shows the need for urgent additional steps to protect the entire watershed area.

The Neretva River project, the subject of an article by a specialist from the University of Split, Croatia, is of considerable socio-economic and environmental importance to Bosnia-Herzegovina and Croatia. The author suggests that a co-ordinated approach is required to resolve issues such as pollution of the estuary and sea, flood protection, water supply, and the management of estuary and marine areas under special protection, among others. Achievement of an integrated plan for water resources management and the sustainable development of river basin and coastal areas depend primarily on sufficient data, i.e. an overall monitoring system.

The coastal areas of Turkey comprise 8,300 km of coastline, which is among the longer coastlines in Europe. Half of its population of 65 million lives in the coastal area, which is heavily polluted by tourism and industries. An improper and insufficient infrastructure for protecting the drinking water supply and for wastewater collection and disposal presents the most serious problems. The authors, who are specialists from Istanbul Technical University and the Institute of Marine Sciences in Urla-Izmir, review studies of the Aegean and Mediterranean coasts, which have been conducted on coastal zone management since the 1980s.

4. The Status of Water Resource Monitoring in the Balkans

In Part III participants from all of the Balkan countries review their country's current water resource monitoring policies and practices. In response to pre-Workshop requests, they emphasise issues particularly related to international waters, providing information on ministerial and other responsible parties, and the agencies that develop and enforce water quality standards. Each country was also asked to describe the extent of

4

transboundary monitoring, and to itemise relevant monitoring sites and parameters measured. They also identify the scientific disciplines responsible for the development of standards and monitoring programmes, and the use of science to evaluate monitoring results.

The unique collection of data and information provided by the authors offer a good starting point for comparative and other purposes, The data will need to be updated from time to time, and this should be a responsibility of the new International Centre.

The countries may be grouped regionally, in accordance with shared river basins, seas, or lakes. They may also be grouped historically; several are now in transition, as they adjust to new democratic and economic systems. National problems with water quantity and quality issues vary, but all report the need to improve water treatment and supply systems. As the case studies show, some countries have already been aware of the need for special transboundary programmes and have worked across borders to establish them. The impact of war on the former Yugoslavia is noted. Bosnia has yet to repair devastated water resources and establish the legal system required to protect them. The Federal Republic of Yugoslavia should be involved in any integrated Balkan approach to serious transboundary issues.

The authors of these articles have adopted a scientific approach and paid close attention to the need for accuracy and precision in monitoring transboundary waters. Administrative information in their work will need to be updated, but the framework they have followed enhances the ability of all regional and international specialists to determine and achieve optimal water management goals.

5. Support for a Balkan Water Network

The five articles in Part IV are written by international water resources experts with different but complementary approaches to the development and practice of optimal international water resource management. They include descriptions of technical assistance needed for improved water resource management, which can be provided by EUROWATERNET, the comprehensive all-Europe monitoring and reporting network established under the guidance of the European Environment Agency, and also the adaptation of systems developed by US agencies for monitoring use in all parts of the country. Both of these articles emphasise the need to retain flexibility as well as accuracy in monitoring systems. All countries need to stay on the cutting edge of electronic networking systems. The authors demonstrate the awareness of this concept and its use in two continents.

A multi-disciplinary, global/local approach is recommended as the basis for supporting a training programme for an International Water-Environment Centre for the Balkans. The need for such a centre is detailed, and possible benefits to be reaped examined. Suggestions are made as to how the centre might be formally established. A second French expert sees the need for such for an international centre from a somewhat different perspective. He presents plans for networking among the countries in the future, with the goal of improving and protecting the endangered water resources of the region. This article explores the need for such an approach and suggests way in which it might be achieved over the long term.

Specialists from Colorado State University in the US and Aristotle University in Thessaloniki write about the effectiveness of a new approach for the resolution of regional transboundary water management issues. They emphasise the need for negotiation to reconcile cultural and historical differences, the calculation of total costs in water resource development, and the incorporation of social and environmental concerns into planning development and use. These broad problems emphasise the need for "hydrodiplomacy," for sustained negotiations that include third party expertise and dialogue based on factual information.

6. Conclusions

The final article presents an overview of the work of the drafting committee and the Workshop teams that supported its conclusions. The main and very significant achievement of the ARW was the fact that all participants reached a consensus on the specific components of the network to be established among all ten countries. This International Network of Water-Environment Centres of the Balkans (INWEB) is designed to contribute to the improved management of the region's water resources. It will develop common projects to train technical staffs, provide access to information and data and sustain a network of experts.

SHARING TRANSBOUNDARY WATER RESOURCES:
THE ROLE OF REGIONAL PARTNERSHIPS IN THE BALKANS

J. GANOULIS
Laboratory of Hydraulics
Department of Civil Engineering
Aristotle University of Thessaloniki
54006 Thessaloniki, Greece

1. Introduction

The importance of transboundary water resources on a global scale is far from negligible: According to reports submitted to the United Nations (UN), about 50% of land on Earth (excluding Antarctica) is located in internationally shared water catchments. About 40% of the world's population live in these areas, which extend over more than 200 international river basins.

Historically, rivers and lakes have been used to determine frontiers between countries. This is why they have been the scene of numerous conflicts throughout history (e.g. the Rhine between France and Germany, the Rio Grande between the USA and Mexico, the Odder and Neisse between Germany and Poland, and the Amur and Ussuri between Russia and China). In many cases river basin boundaries do not coincide with national political borders. Issues and problems of transboundary water management emerge, especially when countries occupy only part of the upstream or downstream area of the river catchment. Water resource sharing then increases in complexity (e.g. the Nile between Egypt and the Sudan, the Middle East conflict over the River Jordan, the Danube River shared many European countries, the Elbe between the Czech Republic and Germany).

A basic question is how, and through what kind of processes water in transborder regions may unify rather than divide sharing nations, and how stakeholders in international water catchments may increase their benefits without causing losses to others. The issue is complex because from technical to ecological considerations, political issues of domestic and external policy are involved.

Using the Balkans as an example, different water interdependencies can be illustrated. In the case of the Evros/Maritza River (shared by Bulgaria, Greece and Turkey) there are no major water supply problems as there are no other water uses besides irrigation. However, complex issues of co-operation exist when it comes to protecting riparian areas from floods and inundation's, as happened recently in February and May 1998. Ecological considerations of the Evros River delta have also become

7

J. Ganoulis et al. (eds.), Transboundary Water Resources in the Balkans, 7–12.
© 2000 *Kluwer Academic Publishers.*

very important in recent years. Albania has protested to the construction of a large dam on the Greek side of the Aoos River, which is shared by Greece and Albania. The number of conflicts on water resources management between Greece and the Former Yugoslav Republic of Macedonia (FYROM) have increased since 1965, due to the intensive irrigation, the plans for constructing new dams in FYROM, and the accelerating pollution of the Axios/Vardar River.

The Nestos/Mesta River, shared by Greece and Bulgaria, presents the greatest challenge in the region. Despite earlier agreements, Bulgaria has been withholding water for increased agricultural and industrial needs. Since 1975 the Nestos flow declined from 1500 million m^3 to 600 million m^3 resulting in repeated Greek protests. A series of negotiations has not resulted in agreement; and this failure is a source of conflict between the two countries. More recently, noticeable pollution from the Bulgarian part has raised the level of tension in a region of Greece highly dependent on irrigated agriculture.

This introductory paper firstly reviews the complexity of transboundary water resources management and different possible strategies for regional negotiations. Secondly, the technical and institutional approaches that may lead to agreements on sharing waters are analysed. Effective implementation of such treaties may be implemented following a bottom-up approach, based on "regional partnerships". This concept is further explained, and takes a concrete form for the Balkans with the creation of the International Network of Water-Environment Centres for the Balkans (INWEB).

2. Complexity of Transboundary Water Resources Management

Within national borders, management of water as a resource involves a number of *internal issues*. These are usually independent of transboundary issues and are the result of physical and institutional characteristics of water resources. The most important of these may be listed as follows:

- Disparities between regions.
- Fluctuations in seasonal and longer time scales.
- Inequality between needs and supply.
- Conflicts in use between different sectors (water supply, agriculture and industry).
- Institutional, legal, economic and social factors.

When transborder water resources are shared between riparian countries, a number of *external issues* should be added to the above, such as

- Differences in political, social, and institutional structures.
- Different objectives, benefits and economic instruments.
- International relations, regulation and conflicts.

Water management becomes even more complex in cases of extreme water events, i.e. floods and droughts. Floods can cause devastating economic and human loss, as happened recently in China, Central America and various parts of Europe. Droughts may result in diachronic water crises due to insufficient water for irrigation, water supply and other water uses. These situations are frequent in semi-arid climates, for example in the Mediterranean region, and may cause substantial socio-economic crises. Floods and droughts are even more difficult to handle in transboundary regions, mainly because of institutional issues.

3. Regional Negotiation Strategies

Many different negotiation strategies are available to modify the complex framework of transboundary water management issues. Decision-makers and those who may negotiate on their behalf have a choice of six universal negotiation strategies:

1. "Win-Win" solutions or Positive sum benefits.
2. "Lose-Lose" solutions or Negative sum benefits.
3. "Win-Lose" negotiations or Zero-sum benefits.
4. Unilateral creation of new facts.
5. Conflict and threats of violence.
6. No action, causing opportunity costs from neglect and/or delay.

The choice of negotiation technique is always subject to political considerations and controversy. Preferences depend on the balance of power among transboundary stakeholders and the cost of concessions. The more powerful and wealthy stakeholders can resort to the creation of facts with minimal risk of counteraction from weaker and impoverished neighbours. They also can afford to make gestures of friendship through "Win-Lose" agreements in the interest of enhancing regional stability.

It can be mathematically proven that "Win-Win" agreements result in positive benefits for both parties and consist of the best trade-off alternative solution. The well-known "Prisoner's Dilemma" shows that failure between interested parties to reach agreement may increase the benefits to one party or the other, but tends to decrease the total benefits. This is because each party left to its own devices will tend to over-use the resource. Co-operation schemes may provide better net benefits to both parties.

However, "Win-Win" solutions may not be always be sufficient when naturally limited water resources are under consideration. In these cases regional networks between water stakeholders can play a very important role.

4. Technical and Institutional Approaches

In order to analyse and understand the origin of water-related conflicts and to provide the "optimal" or "acceptable" or "most beneficial" solution for all parties involved, various approaches have been developed by different communities, including people involved in science, engineering, law, economics, political and social sciences, local communities, administration and policy making. In terms of managing transboundary water resources, the different approaches may be categorised in two groups:

- Descriptive or process theories and models.
- Quantitative or outcome theories and models.

The *first approach* comes mainly from law experts and political analysts, who focus on describing the anatomy of a given situation of conflict or co-operation. They determine the function of different parameters and factors influencing the behaviour of each country, such as the political perception of the importance of water, the international image and status of the country and also economic and institutional issues. Such models, including the behaviour of institutional structures, international negotiation strategies, alternative dispute resolutions and political models are very important. They are mainly prescriptive and not predictive. They do not necessarily give a quantitative output (such as costs and benefits), but they are extremely important for understanding the processes and for analysing the origin and the evolution of conflicts or co-operation.

The *second approach* has been developed mainly by engineers and management experts. Depending on the number of objectives and decision-makers, models may be formulated as *optimisation, multi-objective trade-off* computerised codes, or on the basis of the *team and game theories*. Most of these models are based on the fundamental economic notion of Pareto optimality and are predictive in the sense that they suggest a quantitative "optimal" situation, which should be to terminate a conflict by an equitable resolution between the interested countries. Recent advances and related theoretical developments in this area are documented, and include the application of the fuzzy set theory. However, the success in practice of this kind of *engineering* or *rational modelling* is mainly dependent upon the acceptance between interested actors and countries of the model assumptions, which rely on a set of prescribed objectives and the relative weights or preferences between conflicting goals. In the real world this is not usually the case, and therefore, there is a need to develop better, easier-to-use, interactive and reliable predictive models for transboundary water resources management.

Valuable lessons can be learned from current practices, international conflicts and agreements, which take place throughout the world in international river basins and regional seas. Real life case studies are well documented. Both international freshwater resources in rivers and lakes, such as the Danube, the Rio Grande and the Great Lakes, and coastal waters in regional seas, such as the Baltic Sea and the Black Sea, are included.

By combining the expertise and state-of-the-art knowledge of different scientific communities and disciplines, such as engineering, geography, environmental and social sciences, regional partnerships may contribute to the development of new methods, theories and models in order to resolve more efficiently conflicts and controversial issues in the management of transboundary water resources.

5. Regional Partnerships

Mechanisms and attempts to resolve conflicts over internationally shared water resources are mainly based on:

- International treaties and global declarations.
- Bilateral negotiations and agreements.

International treaties and bilateral negotiations are very difficult to implement, to say the least. Competition over resources, jurisdiction, sovereignty and other rights, historical reasons, complexity of regional issues and lack of participation of involved stakeholders are some of the reasons for the difficulties of the existing system.

"Regional Partnerships" are introduced as a concept and process to mobilise and integrate technical, institutional, social, economic and environmental groups of stakeholders on a regional scale. This is a direct, open and innovative framework for co-operation between riparian scientific communities, water related stakeholders, journalists and politicians in order to build alternative water management mechanisms in transboundary regions. Co-operation between institutions located in riparian counties involved, increased public awareness, involvement, participation and mobilisation are some of the key attributes of "Regional Partnerships".

The main elements of "Regional Partnerships" are:

- Explaining the advantages of "win-win" strategies.
- Establishing common preferences.
- Developing decentralised alliances.
- Reinforcing public education and awareness.

6. Conclusions

Although several hundred international conventions on sharing transboundary waters are theoretically operational around the world, their effective use to achieve the goals they have established is in question. The main reasons for this unfortunate state of affairs are:

- Centralised institutional decision making structures fail to reflect the interests of local water stakeholders.
- The technocratic character of water resources management encourages structural solutions. More often the decisions of building engineering water supply works prevail over water demand management practices.
- The absence of an integrated approach. Water resources management should not only reflect technical but also environmental, economic and social needs.

Regional partnerships are essential for the long-term application and effective implementation of agreements for transboundary water resources management. In this bottom-up approach for decision making, increasing public awareness and public participation are the main components for supporting human networks dealing with water related issues.

Education in primary and secondary schools about water and the environment and continuing education are the main tools for changing cultural attitudes on water use, for introducing new practices for sustainable water management and for increasing citizen responsibility for the resolution of environmental and water related issues. The decision taken unanimously by all participants in this workshop to create an international non-profit association called the International Network of Water-Environment Centres for the Balkans (INWEB) is a decisive step towards that objective.

7. References

1. Ganoulis, J., et al. (eds.), (1996) *Transboundary Water Resources Management: Institutional and Engineering Approaches.* NATO ASI SERIES, Partnership Sub-Series 2. Environment, Vol.7, Springer-Verlag, Heidelberg, Germany.
2. Ganoulis, J. (ed.), (1991) *Water Resources Engineering Risk Assessment*, NATO ASI SERIES, Ecological Subseries, Vol. G29 Springer-Verlag, Heidelberg, Germany.
3. Murphy, I. L. (ed.), (1997) *Protecting Danube River Basin Resources*, NATO ASI SERIES, Partnership Sub-Series 2. Environment, Vol.24, Springer-Verlag, Heidelberg, Germany.

ELECTRONIC NETWORKING - ESSENTIAL TO IMPROVED TRANSBOUNDARY WATER MANAGEMENT IN THE BALKANS

I. L. MURPHY
Consultant
2005 37ᵗʰ St NW
Washington, DC, USA

M. BRILLY
University of Ljubljana
Hajdrihova 28
1000 Ljubljana, Slovenia

1. Introduction

Protection and wise use of water resources should be a top priority for any country, in any part of the world. Continuous access to clean water can help build healthy economies and stable political systems. However, individual countries can only go so far; co-operation among countries is required for the protection of water resource systems that cross national boundaries. A co-ordinated international approach is essential for optimal care and use of vital resources. This is especially true for countries that, like the Balkan countries, are part of an interdependent hydrological region, linked, now and forever, by an extensive system of rivers, lakes, and seas.

Damage has been inflicted on this vital resource by war and by neglect. To restore and provide for optimal resource management, as the experts gathered at this workshop were aware, the countries need to apply state-of-the-art technology, identify special infrastructure needs and, above all, have the support of government leaders, who have been sensitised and made fully aware of the full dimensions and complexities of the problem.

The initiatives begun at this workshop build on work undertaken in other co-operative systems, in the Danube Basin, the Black Sea and elsewhere to achieve the ultimate goal of restoring and improving badly needed fresh water supply systems. Water specialists in ministries and research centres in each country have agreed to plan exchanges of data and information about the quality of the extensive river systems, lakes and seas they share, vital resources which they acknowledge belong to the region and not to any one country.

The main focus lies on a basic component of international water resource management: the monitoring of the quality and quantity of shared aquatic systems. Without the timely exchange of accurate hydrologic and water quality data, it is

13

J. Ganoulis et al. (eds.), Transboundary Water Resources in the Balkans, 13–19.

14

impossible to provide adequately for optimal and equitable use of water resource In order that the best possible results be achieved from this workshop, the need for water monitoring, its use in Europe and in the Balkan countries, and the goals for water management need to be discussed.

2. Standardised Water Data

2.1 WATER QUANTITY MONITORING

The climate in the Balkans ranges from humid to arid due to the distribution of precipitation (Figure 1).

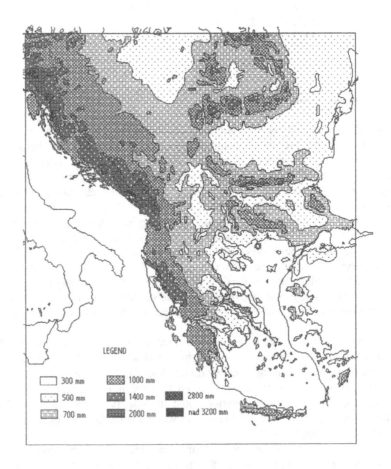

Figure 1. Mean annual precipitation in the Balkan region [5].

Yearly average precipitation is more than 2,000 mm in the mountainous areas from the Alps in the west and the Dinaric mountain and Pindo Mountain in the south. Precipitation reaches the maximum European measurement with more than 6,000 mm in the Montenegrin village of Crkvice near Boka Kotorska. The central part of the Balkan peninsula, from the Pannonian plain in the north to the Thessalia plain in the south is semi-arid, with less than 600 mm of precipitation a year. The Balkan and the Rodopi Mountains in the central peninsula have more than 1,000 mm per year. The eastern part on the Black Sea coast is semi-arid.

Almost half of the water from the mountainous western area disappears underground in karst formations and flows in the shortest direction to the Adriatic. The rest of the water drains via the Danube River towards the Black Sea. Numerous rivers drain the southern part of the peninsula towards the Ionian and Aegean Seas.

The Danube Commission published instructions for hydrological observations and measurements in 1975. Recommended protocol for daily observations includes water level, water and air temperature, ice, and water samples for suspended load, bed load and water quality. Periodic measurements are to be performed seasonally, two to ten times a year, in each of established hydrometric cross-sections. Topography, the water velocity field, the suspended load concentration field and the bed load were included.

Observations of water level and other phenomena began to be recorded in the late 19th century as cited in [1]. Data about discharge, surface and ground water level have been collected following the guidance of the World Meteorological Organisation (WMO) and published annually in formal socialist Balkan countries. Since the recent wars in the former Yugoslavia, measurement of water quantity and data publication have decreased in some Balkan countries.

2.2 STANDARDISATION OF WATER QUALITY DATA

For a number of reasons the development and use of water quality measurements have lagged behind recording and reporting of hydrological conditions. "A wish came true: almost comparable quality data" reads a headline in the most recent edition of the *Danube Watch*, the magazine of the Environmental Programme for the Danube River Basin (EPDRB or Danube Programme) [2]. The article provides a brief history of the EPDRB's development of a monitoring network. Problems with uniformity, from the number of parameters measured at each of the monitoring stations to sampling and monitoring procedures, were not easily resolved.

A joint ad-hoc technical working group, combining both the Danube and Black Sea Programmes, acknowledged that there is as yet "no uniform analytical quality control scheme applied, and the methods by which in-stream loads are monitored for and assessed" are not yet comparable. Any comparative assessment of loads reaching the seas, therefore, is as yet preliminary [3]. Comparable and reliable procedures for data assessment and monitoring have to be found and be in place before it is possible to set clear numeric targets for pollution reduction.

Specialists from all the Danube countries have worked hard to make trend data available in a guide to water resource management for each country and for the basin as a whole. The Programme's Transnational Monitoring Network (TNMN) national experts have established over 70 sampling stations in participating countries. While methodology is not completely compatible, "a common understanding ... has been achieved, although the equipment and methods used for sampling, laboratory analysis and data processing are very different," according to an Austrian expert [3]. The TNMN is intended to strengthen the existing network of transboundary monitoring sites set up by the Bucharest Declaration in 1988. Expert groups in the programme decided parameters, including the list of determinants, sampling frequencies, and analytical procedures.

The results of the TNMN's first year of operation (1996) will soon be published in its first yearbook. These published data will cover the main physical, chemical and biological water quality determinants. The yearbooks will permit a review of trend data for improved management of the Danube. The editors caution that since full comparability has not yet been reached, all data must be looked at carefully. There are changes, for example, in the flow regime, substrate composition and mean water temperature from the source to the mouth of the river that must be taken into account.

Standardisation of monitoring will become more widespread in the Danube Basin and elsewhere in Europe as the guidelines drawn up by the European Environment Agency (EEA) are put into effect throughout the continent. Dr. Tim Lack worked with the consortium appointed by the European Commission to evaluate the monitoring, laboratory analysis, and information management capabilities of the Danube riparian countries, and which finally recommended the establishment of the TNMN [4]. His contributions to the workshop may be found later in this book.

3. Workshop Goals

3.1 ELECTRONIC NETWORKING

Standardised guidelines for collecting essential water resource data were made available to the workshop participants, who also had the chance to discuss information technology. Advances in hardware and software seem to be continuous and exponential. For example a web site called Hydro Info provides access to data about the daily levels and flow conditions of the Danube. Trend data and charts about the hydrology of the river can be developed from this web site. With the increasing use of basic, universal, measurements of water quality, access to valuable water resource data and information may soon be available throughout Europe and in other parts of the world.

This workshop suggested models of electronic networks complementary to the Danube's TNMN and the EEA EUROWATERNET. Beginning with basics: the first need is for the establishment of standardised monitoring procedures across national boundaries at representative transboundary sites in river systems, lakes and seas. The long-term goal will be to agree on procedures that are essential to the equitable, optimal development and use of shared water resources. As the experts working with the Danube Programme frankly admit, without accurate, compatible trend data you cannot make cost-effective, long-term plans for the effective management of a shared water resource.

Information about current transboundary monitoring practices in each Balkan country and about special projects in international rivers, lakes, and seas is included in this book, as well as information about existing monitoring networks. The essential tools for transboundary networking are therefore to some degree already in place. No one is suggesting that full and complete data be available on the Web. Specialists can record and communicate on the "Intranet," using passwords. Any public or private institute conducting the monitoring can use discretion in the release of data. Some Balkan countries already provide periodic summary data about water quality. These could be released on a monthly or yearly basis, or even more frequently in situations of public health or other concern.

3.2 ENCOURAGING NETWORK SUPPORT AND ITS USE

The working groups at this meeting were able to strengthen existing international agreements or, in some cases, initiate informal monitoring and data sharing practices. Long-term goals for using the networks should be kept in mind whatever the level of formality. They need to become available as tools for those responsible for the future development and use of rivers, lakes and seas. All the Balkan countries at the workshop reported that government decision-making bodies use scientific expertise in the formation of water resource policies. This is reassuring. With the help of the EEA, the adoption of uniform quality standards and monitoring will be facilitated and, in effect, mandatory. Networks will capitalise on that development and establish a firm foundation for equitable international water allocation and use policies in the future.

A number of activities can help to support long-term goals:

- Simulation models can be developed that could assist decision-makers in the region in policy formulation, planning, and negotiations.

- Open forums (workshops, conferences and meetings) may be held to discuss all aspects of Balkan water issues at national, regional and international levels.

- Institutes and research centres can help to promote integrated water management concepts and require that published data be supported by metadata.

- Discussion and training modules can be developed to increase the participation of water planners, distributors, and users in decision making about water management.

- Non-governmental organisations representing environmental, commercial and other concerns can be encouraged to participate in the water management process.

- Research and development should encouraged in the region, aimed at transferring new water technologies.

- Training sessions in the fields of water and environmental economics can be initiated.

3.3 BUILDING WEB SITES

Web sites providing public access to information about established international water resource commissions and agencies are increasingly in use. Most are limited to "public relations" information about the agency, such as the purpose, the structure of the co-ordinating organisation and the text of the agreement, and give maps and contact information. Electronic technology allows more informative presentations to be given. The use of relational databases can provide responses to on-line queries and on-line thematic mapping. This is changing the focus of the Internet sites for many organisations from "public relations" to information sharing and exchange.

Web sites can be used to define current management issues for the public and provide a mechanism for the public to ask questions and receive information from the agencies. For example, many web sites provide a facility where users can ask a question via electronic mail (e-mail) and receive an e-mail response to their question. Often criticism, and ultimately opposition to water management decisions, results from a lack of information about and understanding of relevant issues. A more proactive approach is to use Web based on-line surveys and discussion groups to actively solicit public opinion on a potential action.

Web sites can provide a framework for information exchange with water organisations within and between countries in the basin. This is becoming more common with the increased use of on-line databases. For example, in the USA, hydrological and meteorological data are measured by governmental agencies (USGS and NOAA) and made publicly available on the Web. Therefore, all water management organisations in the USA have access to a common database. If organisations want to share certain data, the Web provides a mechanism to accomplish this very rapidly. For example, a specific database can be queried and the records can be easily downloaded to a user's computer for analysis. An important benefit of sharing data this way is that the various agencies can see how similar organisations design their Web sites, what data they collect, the location and frequency of collection and the measurement methods used. Having this kind of information facilitates dialogue to improve the "standardisation" of data collection and display. Ultimately this may lead to the development of national "templates" to facilitate the sharing of data among countries.

Increasingly, texts and summaries of national water related legislation are available on the Web. These provide examples for countries that may be in the process of implementing or modifying legislation. Sharing legislation among the countries in a river basin, or using examples of legislation throughout the world, should enhance the development of international agreements that relate to the river basin as a whole. Prior to the Internet, the timely location of this type of information was often difficult.

As with any new paradigm that rapidly changes conventional approaches, there are cautions and potential dangers that must be considered. Many of these problems existed prior to the Internet, however the speed at which data can be made available on the Internet can exacerbate some problems. There is of course a danger of posting incorrect data to the Web site. While this danger is inherent with any method of documenting data, available technology makes it feasible to post "real-time" information via automatic data publishing software. This minimises the opportunity for quality assurance of the data and it is important to post appropriate warnings to users of such real-time information. Likewise, there is a danger of having censored or altered data put on the Internet either intentionally or by electronic sabotage. Procedures need to be established to provide security of the information that is contained on the web site. Web sites need to clearly define the parameters, units and details of measurement to minimise the danger of any misinterpretation This is particularly important with respect to water quality and ecological data.

4. Conclusions

Most Balkan countries are presently part of, or have access to, established electronic networks designed to improve transboundary water resource management. Such networks have neither been able to provide optimal help for the resolution of transboundary issues nor to help improve water quality at the national or regional level for lack of compatible trend data. With the adoption of EEA standards and possible linkages with the Danube Programme's TNMN programme, this situation will change. Decision-makers in each country will be able to build on these initiatives to establish linkages among countries concerning transboundary rivers, lakes, and coastal waters, so that long-term planning can produce cost-effective treatment and development policies. Firm commitments on the part of research institutes and government ministries can accomplish a great deal towards building regional support for their goals. A regional centre should be adopted to encourage and co-ordinate their networking activities.

5. References

1. Tuinea, D. M. (1998) From Data Acquisition to Data Bank, in XIX Conference of the Danube Countries, Osijek, Croatia, 15-19 June 1998.
2. "A Wish Came True: Almost Comparable Quality Data," Danube Watch, No. 2 1999, Danube Programme Co-ordination Unit, Vienna International Centre, Vienna, Austria.
3. Op. cit., The Black Sea Needs Co-operation.
4. Lack, T. (1997) The work of the monitoring, laboratory and information management sub-group, in Murphy, I. L., Protecting Danube Basin Resources: Ensuring Access to Water Quality Data and Information, Dordrecht: Kluwer, pp. 127-134.
5. Steinhauser F. (1970) Climatic atlas of Europe, WMO, UNESCO, Cartographia, Budapest.

Part II: Case Studies

THE SAVA RIVER

M. BRILLY
University of Ljubljana
Faculty of Civil Engineering and Survey
Ljubljana, Slovenia

T. KUPUSOVIĆ
Hydro-Engineering Institute
Faculty of Civil Engineering
University of Sarajevo
Sarajevo, Bosnia and Herzegovina

O. BONACCI
University of Split
Faculty of Civil Engineering
Split, Croatia

D. LJUBISAVLJEVIĆ
University of Belgrade
Faculty of Civil Engineering
Belgrade, Yugoslavia

1. Introduction

The Sava River, a tributary of the Danube River, was the largest national river basin in the former Yugoslavia and had a population of seven million [6]. A small part of the basin has always belonged to Albania. The basin is now international, its catchment area providing drainage for four countries: Slovenia occupies 12% of its total watershed area, Croatia, 16%, Bosnia and Herzegovina (BIH), 40%, and the Federal Republic (FR) of Yugoslavia 32%. The Sava River and its tributaries are both boundary and transboundary rivers. Some national co-ordinated water management systems and construction projects are now divided between the independent states.

Water management in the former Yugoslavia was the responsibility of the six republics; the federal government acting as the co-ordination of water management activities The republics' hydrometeorological institutes measured both water quantity and water quality. The data were collected and sent to a federal hydrometeorological institute, which then published them in yearbooks. This institute was responsible for co-ordination on a national level and for co-operation with foreign countries.

J. Ganoulis et al. (eds.), Transboundary Water Resources in the Balkans, 21–31.
© 2000 *Kluwer Academic Publishers.*

UN experts [1] completed a water management study of the Sava River in 1972. The study was never accepted as a water management plan. In the 1970s, however, co-ordinated action was undertaken for a comprehensive study of the basin. By 1988 major components of the study had been completed including an overview of monitoring and flood protection needs. After it began operation in 1994, the monitoring, laboratory and information management subgroup of the Environmental Programme of the Danube River Basin (Danube Programme) established five monitoring border and transborder stations on the Sava, one on the border between Slovenia and Croatia and four on the border between Croatia and BIH.
(http://www.icpdr.org/DANUBIS/, http://www.rec.org/danubepcu)

2. Water Balance

The Sava River basin covers 96,400 km ^{2}, 25% is karst, primarily in the upstream part of the watershed that features caves and underground rivers (see Figures 1, 5 and 4). The western and southern parts of the watershed are a large mountainous areas with more than 1,000 mm of precipitation per year. The northern part of the basin forms part of the Panonian Plain, which is a less humid region. The southwestern watershed border in the karst region is quite complex, with water flowing in different directions in response to hydro-geological conditions.

The mean annual discharge of the river Danube at the city of Belgrade is about 1,700 m ^{3} per second. The Sava River contributes about 25% of the total discharge of the Danube River, from 15% of the total Danube River basin watershed [2]. The ratio of the highest to the lowest monthly average discharge is 9.6:1.

The main stream extends from west to east and is 946 km in length. The river changes with respect to slope 10 km downstream of Zagreb. In the upstream parts, and indeed right up to its source, it has a gravel bed with a slope of more than 70 cm per km, whereas downstream it has a silt bed, is 680 km in length with meanders, and has a slope of about 4 cm per km and a large area of wetlands. Tributaries provide the river with a large amount of water.

The border rivers between Slovenia and Croatia include a tributary from the west, the canalised Sotla (Sutla) River and a tributary from the east, the Kolpa (Kupa) River with its tributary Čabranka. The Kolpa River has a watershed of 9,800 km^2 and a mean annual discharge of 280 m^3 per sec [3]. Part of the watershed is covered with untouched natural forest.

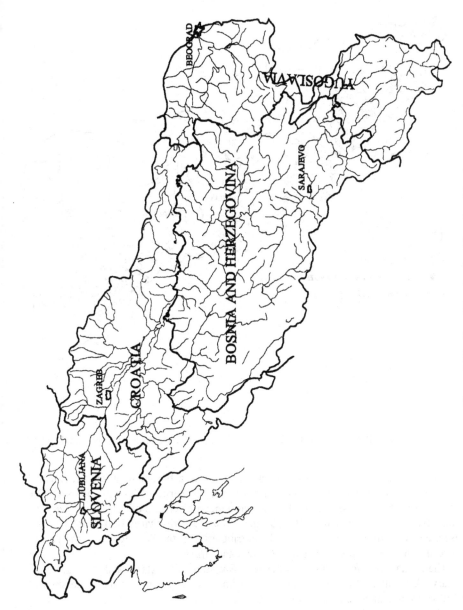

Figure 1. The Sava River basin with national borders.

Figure 2. The Sava River basin with main sub watersheds, heritage protected areas and karst.

The Una River, a right bank tributary that forms a border between Croatia and B&H, has a watershed of 9,640. km^2 and a mean discharge of 290 m^3 per sec. The Vrbas River with a watershed of 6,386 km^2 has a mean discharge of 100 m^3 per sec. The Bosna River has a watershed of 10,460 km^2 and a mean discharge of 170 m^3 per sec. These are rivers within B&H and right-bank tributaries of the Sava River on the part of the stream that forms a border between Croatia and B&H.

The Drina River forms a border between B&H and Yugoslavia and is a right-bank tributary. A small part of its watershed (160 km^2) is in Albania. It has watershed of 19,570 km^2 and a mean discharge of 370 m^3 per second.

The average daily minimum discharge of the Sava River at its mouth was 390 m^3 per second for period 1931 to 1970 [1]. The lowest observed flow was 194 m^3 per second in the year 1946.

3. Water Quality

3.1 HISTORICAL BACKGROUND

The monitoring of the water quality of the Sava River began in 1965. Water is divided into four quality classes according to its use. The assessment of the category is based on the classification system used for chemical, microbiological and biological analyses and was approved by the then Federal Government in 1978. The republics of the former Yugoslavia were responsible for monitoring, with each of them using different methods. Primarily, monitoring was the responsibility of each republic's hydrometeorological institute and each one used slightly different procedures for sampling, chemical analyses and water classification. The water quality classes of the Sava River and its tributaries are represented in Figure 2. The water was classified according to mean annual suprobic indexes published in 1985.

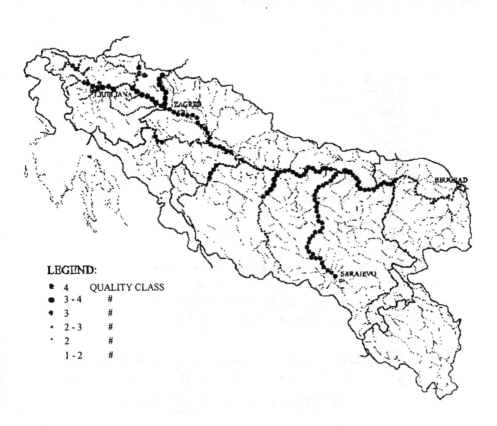

LEGEND:

♠	4	QUALITY CLASS
●	3 - 4	#
♦	3	#
•	2 - 3	#
·	2	#
	1 - 2	#

Figure 3. Water quality of the Sava River and its tributaries.

The institute for water management [1] completed a study for the development of a common monitoring structure. There were 15 automatic stations, 19 stations sampling three times per week, and 96 stations sampling from six to twenty times per week. The proposals for harmonised procedures were also developed using international state of the art technology, primarily based on examples from the United States.

3.2. RECENT INVESTIGATIONS

Recent investigations show that the Sava River has a relatively high BOD content in comparison with the Danube River and its other tributaries, see Figure 4, [2]. BOD in the Sava River is the lowest of any large Danube tributary and decreases the BOD level of the Danube itself at their confluence. The Transnational Monitoring Network (TNMN), a project of the Environmental Programme for the Danube River Basin, has established seven stations in the Sava River basin. The monitoring network was established recently and data have not yet been made available.
(http://www.icpdr.org/DANUBIS/, http://www.rec.org/danubepcu).

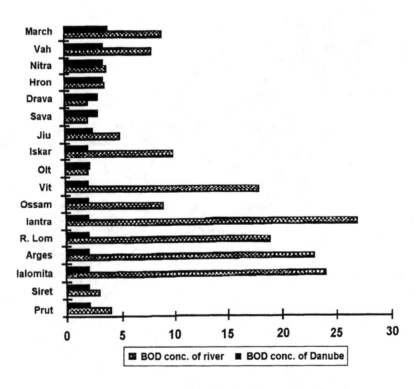

Figure 4. BOD levels of the Danube River and its tributaries.

At its source, Sava waters are in the highest category. Quality gradually decreases downstream averaging a II-III classification on the border between Slovenia and Croatia. On short reaches of the river downstream of Ljubljana, sewerage system outflow water has decreased water quality to class IV levels (Figure 3). Downstream of the Zagreb region to the inflow of the Kupa and Una Rivers, the river is classified as class III. Further downstream water quality is classified as II-III as far as the inflow of the very polluted Bosna River. Inflow from the latter decreases water quality in the Sava River to class III. The inflow of the Drina River improves water quality to the II-III level up to the confluence with the Danube. There are very polluted small creeks in the industrial parts of the Zagreb region and in the heavily polluted Bosna River downstream of the Sarajevo region. The inflow of relatively clean water from the Una and Drina Rivers improves the water quality of the Sava, see Figure 3.

In 1998, water in the Croatian part of the river was classified as being class IV, according to Croatian classification practices. The quality of water in the Sava River is estimated to be class II in Yugoslavia. As the river Sava has recently become a transboundary river, it is very important to establish an appropriate framework for protection of its water quality.

4. Water Management of the Sava River and its Tributaries

The Sava River and its tributaries are important water resources for the development of its basin countries. Care needs to be taken however that development does not conflict with the need for the protection of natural resources, national parks and regions with very high biodiversity.

4.1 WATER SUPPLY

Shallow ground water aquifers are the main source of water supply for numerous settlements along the Sava River, including the capital cities of Ljubljana, Zagreb and Belgrade, which are located on its banks. Belgrade extracts some water from the Sava River and treats it for water supply purposes. Ljubljana can use ground water that does not require treatment. Almost all cities and villages along the river use ground water from alluvial aquifers for water supply without special treatment.

The Belgrade Waterworks Authority secures about 3 m^3/s by abstracting water from the Sava River - more then one-third of the river's capacity. At the treatment plant "Makis", where water from the Sava river is treated, advanced water treatment is applied (preozonation, coagulation-flocculation-settling, ozonation, double layer filtration, and decontamination with chlorine). Due to the use of robust applied technology processes in this treatment plant there have so far been no problems with treated water quality.

4.2 WASTE WATER TREATMENT

There are no wastewater treatment plants on the main stream of the Sava River downstream of Ljubljana. Ljubljana's city council constructed a mechanical treatment plant in 1990 and development of the second stage of the treatment plant is now in progress. The city council of Zagreb contracted in 1999 for the construction of a sewage treatment plant. In the FR of Yugoslavia there are several wastewater treatment plants in the Sava watershed, including those of the cities of Ruma, Indjia and Valjevo. Construction is about to begin of an advanced wastewater treatment plant (with nutrient removal) for the city of Sabac, located on the bank of the Sava. The Belgrade sewerage system consists of four subsystems, with one of them (the smallest subsystem called Ostruznica) forms part of the river Sava watershed. Design projects for Belgrade wastewater treatment plants have been completed, but their construction has not yet begun. Unfortunately, open dumps for municipal and industrial solid waste disposal are presently characteristic of the area.

Sarajevo has the largest sewage treatment plant on the Sava River basin. It includes facilities for biological treatment and was designed to serve a population of 600.000. However, the system has operating problems, because it was not properly constructed as a separate facility. During rainfall inflow was not concentrated enough for proper operation and thus its quantity was too high for the treatment plant, which was also overloaded with sand flushed from land surfaces, resulting in parts of it not being able to function properly. Under normal conditions, the treated water reached the anticipated acceptable levels of quality (12-15 mg/l BOD5). The energy generated by the treatment plant was too low for operational purposes, which meant higher operating costs. During the war the treatment plant did not operate at all.

4.3 RIVER CONSERVATION

The mainstream of the Sava River has been partly channelled over the past hundred years. The project provides for flood protection and the drainage of a wet and arable plain downstream from the Slovenian–Croatian border to the Danube inflow point. There are also large systems of levees along the river. A large diversion channel was constructed to provide flood protection for Zagreb, which was badly damaged by the Sava flood of 1964. About 60 km^2 of a heavily urbanised area and 300 km^2 of the surrounding area were flooded, 45,000 inhabitants were evacuated and 17 people lost their lives. For the protection of cities downstream of Zagreb, a large flood control system has been constructed. The main flow of the Sava River between Ljubljana and Zagreb has decreased almost 2 m in the past twenty years due to the extraction of sediments.

On the downstream part of the river in Yugoslavia a large inundated area is very well protected by embankments for a total length of 771 km. Backwaters of the reservoir Djerdap I (Iron Gate I) on the Danube has slowed the lower course of the river Sava.

4.4 IRRIGATION AND DRAINAGE

In 1965 the lower reaches of the basin suffered a 100-year flood. The flooded area measured 7,890 km^2 and the area flooded yearly was 2,980 km^2. Wetlands cover 90 km^2 and fishponds 50 km.^{2}. There was a water management plan for the protection of the entire area up to the year 2,000 [6]. The drainage works were completed in part. The

Bosut river drainage area includes a flat plain of 2,913 km^2 with a pumping station at the point of outflow to the Sava River. A great part of the drainage area is now in Croatia. The pumping station located in FR of Yugoslavia does not work. A new navigation channel between the Danube River and the Sava River in Croatia could be used for irrigation and drainage purposes.

The Sava River basin has 13,295 km^2 of arable land of which more than 10,000 km^2 could be irrigated. There are, however, no large irrigation facilities along the Sava. Some small systems divert water from the main stream for irrigation purposes. The estimated flow of about 260 m^{3}per second needed for irrigation is almost equal to the low flows of the Sava River in a dry year.

4.5 NAVIGATION

A century ago the Sava River was navigable from Belgrade to Ljubljana, before rail connections were established in the region. Navigation for commercial purposes is now possible as far as Sisak. There are plans to make the river navigable upstream to Zagreb and possibly from Zagreb to Rijeka on the Adriatic Sea, or from Zagreb to Ljubljana and Trieste in Italy. A channel between the Sava River and the Danube River in Croatia is now under consideration. Its construction will require a 560 km traffic corridor between the Danube and the Adriatic. A more realistic approach would be to combine a railway connection of 61.5 km with a Danube-Sava canal 340 km long. The Sava River (on the border with BIH) would include three water locks (Zupanja, Jasenovac, Sisak) and connect with a 160 km double-track railway that would run from Rijeka to Zagreb.

4.6 HYDROPOWER FACILITIES

There are four large and several small hydropower stations in Slovenia, which utilise about 20% of the hydropower of the Sava River watershed [8]. No other power stations have been constructed on the main course of the river downstream of these stations. However, a chain of hydropower plants located upstream of the town Sisak is included in Slovenian and Croatian water management plans. An Austrian/Slovenian consortium is negotiating for the construction of hydropower facilities in Slovenia. Several hydropower plants were constructed on the mountainous part of the tributaries in BIH, but only a small part of their potential energy has been used: Una River (1.7%), Vrbas River (28.5%), Bosna River (2.9%) and Drina River (33%). A plan for the development of more hydropower plants on the Drina River in the former Yugoslavia was dropped when negotiations between the republics failed.

4.7 INDUSTRY AND POWER

A nuclear power plant is situated near the Sava River in Slovenia close to the Croatian border. There are also several thermal power plants in Slovenia and in the FR Yugoslavia near the border with BIH. There are several industrial facilities along the river and on its tributaries; some of them are no longer producing because of the slow economy and the impact of military action during the war. Industrial effluents have heavily polluted the tributaries and some parts of the Sava itself, as shown in Figure 3.

4.8 WETLANDS AND NATURAL CONSERVATION

Springs of the Sava River are located in the Triglav national park (http://www.sigov.si/tnp/) which covers an area of 838 km^2 half of which belong to the Sava River basin. The Plitvice lakes National Park and Risnjak are in Croatia (www.ring.net/duzo) on tributaries of the Sava.

There are several national parks in the FR of Yugoslavia and in BIH. There is also a national park in the upstream part of the Drina watershed. The Drina River basin is characterised by high biodiversity and protected ecosystems. A number of migratory and endemic species are concentrated within the national parks. The Durmitor park also features a glacial lake, the river Tara watershed and canyon, the Biogradska gora park has a glacial lake Biogradsko, and the Tara park includes the river Drina Canyon and several strictly protected bioreserves. There are several protected areas in BIH with autochthonous forests (Perućica, Janj, Plješevica), mountain lakes Planinska jezera (Treskavica) and Prokoško jezero (Vranica) among others.

Large wetland areas are protected in the northern part of the FR of Yugoslavia and special measures are being taken to preserve some wetlands in the inundated area. At the present time there are several protected wetland areas and autochthonous forests on the downstream part of the river Sava watershed in Yugoslavia (Obedska pond, Orlaca, Bosutska forest and Zasavica). There are several natural heritage areas in the Sava basin. Slovenia would like to protect the Kolpa River on its border with Croatia, but this is not part of Croatia's water management plan.

4.9 THE DANUBE - SAVA MULTIPURPOSE CANAL

Croatia has for a long time discussed the Danube-Sava canal as a possible route for future traffic between Central Europe and the Mediterranean region. The canal was featured as a strategic structure for traffic, water management and agriculture for Croatia in the Croatian Parliament's document "Development Strategy of the Republic of Croatia" (1990). The government issued the "Decision on Preparations for the Construction of Multi-Purpose Danube-Sava Canal" in 1991.

The canal will extend 61.5 km from Vukovar on the Danube River to Samac on the Sava. In order to meet navigation requirements in accordance with the UN/ECE 5th class for navigable routes, it is planned to be 34 m wide at the bottom, 58 m wide at the water table and 4 m deep with an average excavation of 10 m The maximum excavation depth is 22 m. There will be two weirs with navigation locks. For most of the time the water level of the Sava River will be higher than that of the Danube River and the water will flow toward the Danube. Even so it will still be possible to reverse the flow.

The canal will drain 2,350 km^2 of lowlands, provide water for irrigation, regulate the water regime according to forest vegetation requirements, thereby enriching the water supply so as to provide environmental protection for the settlements along the canal as well as the possibility of its utilisation for processed water.

5. Conclusion

Maintenance and reconstruction of water structures built in the former Yugoslavia are undecided political questions; their solution provides unique challenges. An appropriate response would provide an opportunity for the development of sound transborder water management, which has not been possible until now. A pilot study is proposed for the Sava River basin, according to the "Convention on the Protection and Use of Transboundary Watercourses and International Lakes" and the EU "Water Framework Directive", which strongly emphasise water management and international co-operation within river basins.

The countries around the Sava River basin are currently going through a transitional post-war recovery period and are in economic crisis. The water quality of the Sava River has improved because of the decrease in industrial production during the recent wars and the slowing down of the economy. Pollution levels have recently increased slightly. Investments in treatment plants have been postponed for the last ten years in the former Yugoslav countries, including Slovenia.

Several transboundary effects have been noted, including a decrease in flow rate due to water use in upstream countries, flooding due to inappropriate operation of detention basins in upstream countries, accidental pollution, frequent oil spills, risk from nuclear pollution, and pollution from petrochemical and metallurgy industry coming from upstream countries. As the river Sava has become a transboundary river it is very important to establish an appropriate framework for the protection of its water quality.

6. References

1. Djukic M. et al (1975) Common methodology for monitoring of water quality of the Sava River basin, Institute for water management, Jaroslav Cerni, Belgrade, Report of the Co-ordination Committee of the Sava River Project.
2. Haskoning, (1994) Danube Integrated Environmental Study, Report phase 1, HASKONING, Royal Dutch Consulting Engineers and Architects, EPDRB report.
3. Institute of Hydraulic Engineering "Jaroslav Cerni" (IHE Jaroslav Cerni), (1956), Water Power Resources of Yugoslavia, Yugoslav National Committee of the World Power Conference, Beograd.
4. Markovic R.D. et al (1969) Hydrological Study of the Sava River, Federal Hydrometeorological Institute, Belgrade, Study report co-ordinated by the Sava River Authority of Croatia.
5. Prohaska S. (1976) Hydrology of The Sava River - 1976, Institute for Water Management Jaroslav Cerni, Belgrade, Report for Co-ordination Committee of The Sava River Project.
6. Regan D. (1969) Yugoslavia - Conservation and regulation The Sava River Basin, Projection to 1985 and 2000 year, Zagreb, Study report co-ordinated by Institute of agricultural economy and sociology University of Zagreb.
7. The Report for the Sava River Authority of Croatia.Study for the regulation and management of the Sava River - Final Report, (1972), Consulting Eng. Consortium Polytecna - Hydroprojekt, Carlo Lotti/Co. Prague - Roma
8. Union of self-self-governing water communities of Slovenia (1978) Water Management Master Plan, co-ordinated by Water management Institute, Ljubljana

THE NESTOS/MESTA RIVER: A TRANSBOUNDARY WATER QUALITY ASSESSMENT

E.PAPACHRISTOU, J. GANOULIS, A. BELLOU,
E. DARAKAS, D. IOANNIDOU
Department of Civil Engineering
Aristotle University of Thessaloniki
Gr-540 06 Thessaloniki, Greece

1. Introduction

The management of transboundary water resources is an important matter for scientists and policy makers. Water use and the protection of water quality and ecosystems are the main aspects of the problem, which should be faced in co-operation with neighbouring countries. This is not always easy, due to different socio-economic and political conditions, therefore the different needs and priorities of the sharing countries should be discussed.

The Nestos River is the most important water resource in its region and has been the object of negotiations between Greece and Bulgaria for many years. Its famous ecosystem is in danger because of the pollution caused by various human activities and large-scale hydraulic works (dams) constructed along the river. The protection and management of Nestos waters are of great economic and ecological importance for both countries and should be based on a thorough transboundary environmental study. Such a study was undertaken by the Division of Hydraulics and Environmental Engineering, Department of Civil Engineering of the Aristotle University of Thessaloniki (AUT) in Cupertino with the Faculty of Hydrotechnics Department of Water Supply, Sewerage, Water and Wastewater Treatment, University of Architecture, Civil Engineering and Geodesy, Sofia, Bulgaria. The study was performed as part of the INTERREG programme, under the topic "Transboundary Co-operation". Scientists of different disciplines from both universities collaborated on the project. It is worth mentioning that this was the first time the two countries carried out a co-ordinated programme for the monitoring of river pollution

2. General Description of the Study Area

The Nestos River rises in the Rila mountains in southern Bulgaria and flows some 230 km through Bulgarian and Greek territory before emptying into the North Aegean Sea. About 100 km of the river flow through Bulgaria and about 130 km through Greece (Figure 1).

J. Ganoulis et al. (eds.), Transboundary Water Resources in the Balkans, 33–40.
© 2000 *Kluwer Academic Publishers.*

34

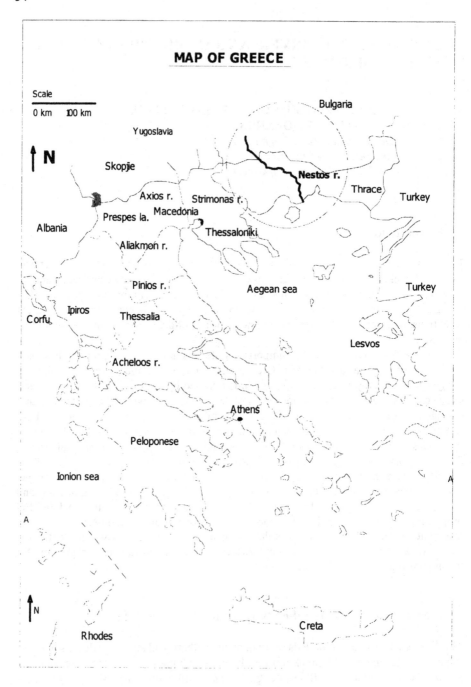

Figure 1 Location of the River Nestos.

The total catchment area of the river is 5749 km^2, of which 2312 km^2 (40%) belong to Greece. The morphology of the area is mountainous with the exception of the delta region, which covers an area of 440 km^2. The climate is Mediterranean and becomes transitional Mediterranean towards the north. According to data of mean monthly flows recorded by the Public Electric Power Corporation (PPC) for the period 1965-1990, only in a very few cases was the maximum discharge over 150 m^3/s, while the minimum discharge was often lower than 10 m^3/s. During recent years three dams have been constructed on the Greek section of the river: (a) Thisavros, (b) Platanovrisi and (c) Temenos. All these dams will provide hydroelectric power generation and water supply for irrigation networks. There has been concern about the impact of these works downstream, particularly in areas of natural beauty and ecological interest, such as the "Tembi of Nestos" and of course the delta area which is protected by the RAMSAR Treaty.

Administratively the whole study area belongs to three prefectures, those of Drama, Xanthi and Kavala.

3. Inventory of Pollution Sources

The pollutant loads discharged into the Greek section of the river are mainly from:

- Incoming pollution from various activities in the Bulgarian area.
- Domestic wastewater from villages discharging into the Nestos River and its tributaries.
- Wastewater from the few industries and handicrafts factories located in the study area.
- Drainage ditches which receive, along with storm water, residues of pesticides and fertilisers.

3.1 URBAN POLLUTION

The Nestos River Basin comprises 82 villages and small towns with a total population of about 35,000 inhabitants. Chrisoupoli is the largest town (7,100 inhabitants) and is the only one served by a central sewerage system combined with a treatment plant. Stavroupoli and five other villages only have sewer networks. Since there are no treatment plants, the Nestos River receives their untreated wastewater.

The rest of the villages are served by private cesspools which are mainly drain tanks, polluting the groundwater and consequently causing indirect pollution to the Nestos River.

3.2 AGRICULTURAL POLLUTION

The agricultural activities of the northern mountainous section of the study area are very limited. There are only 15,660 strems of cultivated land and therefore the pollution is not significant.

The southern part of the study area, starting from Toxotes and extending down to the delta area, has many flat fertile pieces of land with a developed irrigation network. According to data collected from the relevant authorities the cultivated area in Kavala Prefecture is 178,512 strems and in Xanthi Prefecture is 128,792 strems. Pesticides and fertilisers are conveyed to the Nestos via drainage ditches. Agricultural pollution seems to be the most important pollution source in the delta area.

3.3 INDUSTRIAL POLLUTION

Data collected for the study area indicate the following:

- The Prefecture of Drama has no industrial or handicraft activities that could pollute the Nestos River.
- The Prefecture of Xanthi has only a few small handicraft factories that could be a source of pollution for the river; the main receiver of the industrial area is the Laspias stream discharging into the Thracian Sea, east of the Nestos.
- The Prefecture of Kavala has some industries and handicrafts factories, which could pollute the Nestos.

4. Water Quality Parameters

In order to determine the water quality conditions of the Nestos River a programme of *in situ* measurements, samplings and laboratory analyses was carried out. Three sampling points were selected at representative locations along the river, at Potami, Stavroupoli and the delta area (1, 2 and 3 respectively). The first sampling point at Potami is only 7 km downstream from the Greek-Bulgarian border, and since there are no intermediate pollution sources, water quality at that point reflects incoming pollution from Bulgaria. Stavroupoli is the second station and is 47 km from the border, near the middle of the Greek part of the river. From this point the river enters the plains area below the mountains. The last of the three dams constructed by PPC is located just upstream from this site. Water quality data analysis for this station is significant for establishing a database with historical data useful for future environmental impact assessments of the dams. The third sampling point is located at the delta, about 1 km upstream from the estuary. At this point the river has received the total amount of pollution from Bulgaria and Greece, which is reflected in the water quality of the river as it enters the sea.

During the programme sampling sets were collected twice a month, and a total of 39 sets were analysed. The following parameters were determined:

- pH
- Temperature
- Conductivity
- Dissolved Oxygen (D.O.)
- Suspended Solids (S.S.)
- Biochemical Oxygen Demand (BOD_5)

- Chemical Oxygen Demand (COD)
- Phosphates (PO_4^{3-})
- Nitrates (NO_3^-)
- Nitrites (NO_2^-)
- Ammonia (NH_4^+)
- Heavy Metals (Cu, Pb, Cd)

Temperature, pH, conductivity and dissolved oxygen were measured *in situ*. The rest of the parameters were determined at the laboratory of Environmental Eng., Department of Civil Engineering, AUT.

In Figure2 time series plots of basic water quality parameters indicating organic pollution of the Nestos River at the delta station are shown. Similar diagrams have been worked out for all the parameters at all three stations.

The results of all these measurements were then evaluated statistically, and suggest the following conclusions about the water quality parameters.

<u>Temperature</u> increases from the border down to the mouth of the river, varying within normal levels for the area.

<u>pH</u> ranges in normal levels without significant changes between the sampling points. The values meet the standards for drinking water as well as for fish propagation.

<u>Conductivity</u> increases from upstream to downstream, with much higher values at the delta, due to seawater intrusion. This is also the reason that drinking water standards are sometimes exceeded at the delta station.

<u>Dissolved Oxygen (DO)</u> and saturation percentage of DO (DO%) varies along the river with frequent supersaturated values. This is indicative of good water quality with respect to organic pollution, but also of the existence of phytoplankton. Station 2 seems to have the best values with higher D.O concentrations than at the other two stations. The values of both parameters meet the standards for drinking water and fish propagation.

<u>Suspended Solids (SS)</u> at station 3 tend to have lower values than at station 2, and values at station 2 are in turn lower than at station 1. This could be explained either by incoming pollution from Bulgaria or by natural causes, like the decrease of water velocity as the slope of the riverbed gradually decreases. Values at stations 1 and 3 sometimes exceed the guide values for class A1 drinking water, as well as for salmonides and cyprinides propagation.

<u>Biochemical Oxygen Demand (BOD_5)</u> has higher values at station 1, while stations 2 and 3 do not have significant differences. This is attributed to incoming pollution from Bulgaria and gradual assimilation along the river. Nevertheless, values at all three stations exceed the guide values for classes A1 and A2 drinking water and the guide values for salmonides propagation, while at stations 1 and 3 the standards for cyprinides propagation are also exceeded.

<u>Chemical Oxygen Demand (COD)</u> values have no statistical differences among the three stations. Station 1 presents higher extreme values, which might reflect incidences of pollution in Bulgaria. The existence of extreme values at the other two stations shows that the part downstream of Stavroupoli also accepts a pollution load coming from the surrounding watershed. The values of COD along the river are lower than the guide values for class A3 drinking water, in this way meeting the legislative standards.

38

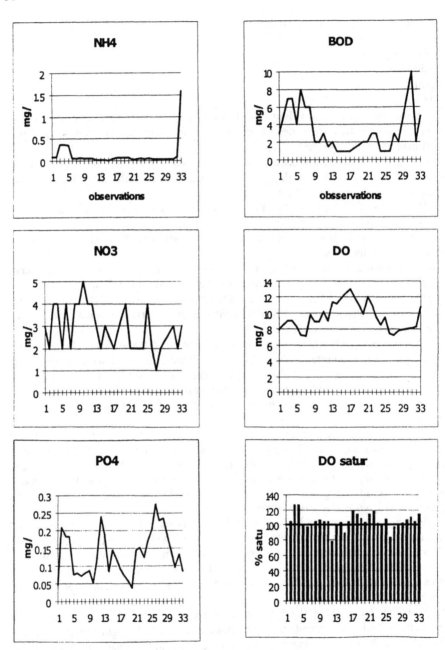

Figure 2. Plots of basic water quality variables in River Nestos at Delta station (6/92-10/93).

Ammonium (NH$_4^+$) is characterised by the high variance at all stations. Stations 1 and 3 have higher values than station 2. Station 3 presents extreme values that occur at different times and are of a different magnitude to station 2 (maximum at station 3 was 1.54 while at station 2 the level was 0.45). The entire above can be attributed to imported pollution (industrial, domestic or agricultural) and to pollution entering downstream at Stavroupoli (domestic or agricultural). Values of NH$_4^+$ show higher levels of water quality than the standards, i.e. the guide values for class A1 drinking water and for salmonides and cyprinides propagation.

Nitrites (NO$_2^-$) Higher values were again measured at station 1, while the other stations did not differ significantly. Imported industrial or urban pollution is probably the explanation. Guide values for salmonides and cyprinides exceed standards at all three stations. At station 1 the values measured are about three times higher than the guide values for cyprinides.

Nitrates (NO$_3^-$) values are of low or moderate levels, with station 1 presenting higher values than station 2, probably due to imported pollution. Station 3 has higher values than station 2, probably due to pollution from the plains downstream of Stavroupoli. However, this parameter meets all the standards for potable water.

Phosphates (PO$_4^{3-}$). The pattern is the same as for nitrates, indicating the same pollution sources in both countries. The measured values meet the guide values for all classes of potable water

Heavy metals (copper, lead, and cadmium). There was a clear declining trend from station 1 to station 3 for all three parameters, indicative of industrial pollution imported from Bulgaria. Copper values exceed guide values for class A1 drinking water at stations 1 and 3, while the values are very close to the limits for fish propagation. Lead meets the standards, yet is very close to the limits. Cadmium exceeds the standards (imperative values) for classes A1, A2 and A3 drinking water at station 1. Concentrations of the three metals were also measured in sediment. They present more or less the same pattern, declining as they move downstream. Unfortunately there are no legislative standards for water uses and consequently no comparison can be made with the measured value.

5. Conclusions

Measurements and analyses of water quality parameters reported by the project proved to be a useful tool for the assessment of river water quality. The following conclusions have been drawn:

- The actual state of water quality of the Nestos River for the period of research was established.
- Differences in water quality at different locations on the river have been clearly distinguished indicating that the river is more polluted at the entrance from Bulgaria. Quality improves gradually towards Stavroupoli, deteriorating again as the river moves towards the delta.
- Comparison with standards set by EU and Greece for various water uses shows that the present state of water quality in general meets standards for class A3 drinking

water. According to the results of water quality analyses, fish propagation is not problematic but needs care, especially for salmonides.

- Different properties of water quality parameters reflect different types of pollution sources e.g. heavy metals concentrations at station 1 were indicative of industrial pollution imported from Bulgaria, while nutrient concentrations at station 3 indicated agricultural pollution. Since pollution can now be attributed to certain geographical areas, other potential causes of pollution may be more easily identified and investigated. Identification of pollution sources may also be made easier by a combination of other tools, such as the use of geographic information systems.

Planning and management of water quality in the river basin needs to be supported by the collection of updated data on a continuous basis. Continuous monitoring of the Nestos River is necessary for this purpose, so that decisions can be taken about compliance with water quality standards and the corrective measures, which may need to be taken. It also has been established that management of a transboundary river needs the close collaboration of the transboundary countries with respect to all the findings of the study.

6. References

1. Papachristou E. at al (1994) "Investigation of the pollution of transboundary, between Greece and Bulgaria, Nestos River". EU, INTERREG program, project final report, Thessaloniki.
2. Argiropoulos D., Ganoulis J. and Papachristou E. (1996) "Water quality assessment of the Greek part of Nestos (Mesta) River" in: Ganoulis J., et. al. (eds) Transboundary Water Resources Management: Institutional and Engineering Approaches. NATO ASI SERIES, Partnership Sub-Series 2. Environment, Vol. 7, Springer-Verlag, Heidelberg, Germany, pp. 427-438.
3. Argiropoulos D., Papachristou E. and Ganoulis J. {1994) "Statistical assessment of water pollution in the Aegean Rivers: the case of Nestos". Sixth Meeting of the Regional Agency for the Environment, Provence – Alpes – Cote D'Azur, France.

THE MONITORING PROGRAMME OF THE LAKE OHRID CONSERVATION PROJECT

Z. SPIRKOVSKI, Z. KRSTANOVSKI
Hydrobiological Institute
96000 Ohrid, FYROM

L. SELFO, M. SANXHAKU, V.I PUKA
Hydrometeorological Institute
Tirana, Albania

1. Introduction

The significance of Lake Ohrid as an unique aquatic ecosystem is well known not only to the people of its border countries of Albania and the Former Yugoslav Republic of Macedonia (FYROM), but also to scientists throughout the world. Its richness in biodiversity and especially in relic and endemic species, as a result of its great age, geographic isolation, and the stability of its ecological conditions, gives it global significance.

The catchment area of the lake has a population of around 131,000 people: 43,000 in Albania and 88,000 in FYROM. Human activities continue to influence lake water quality resulting in changes of habitat conditions, which have disrupted the million-year old, well-maintained dynamic balance of the ecosystem. This has become particularly evident in the last half of this century.

Although certain protective measures have been undertaken, the present state of the lake shows the need for urgent additional measures to accomplish efficient protection of the whole Lake Ohrid watershed area. The transboundary initiative responding to this need for protection of the natural resources of the lake, the Lake Ohrid Conservation Project, is a joint effort of both countries sharing the lake. It consists of four components: Institutional Strengthening, a Monitoring Programme, Watershed Management, and Public Awareness. The major goal of this project is to ensure the restoration and future conservation of the lake. The project is funded by The Global Environmental Facility.

2. Lake Ohrid

The karstic tectonic and oligotrophic Lake Ohrid (41°05'N, 20°45'E; max. length 30.4 km, max. width 14.8 km) is between 4 and 10 million years old, and is one of the oldest lakes in the world, and the very oldest in Europe. It is situated in the Ohrid valley, in the southwestern part of FYROM, on the border with the Republic of Albania (Figures 1 and 2) at 693.17 m above sea level.

J. Ganoulis et al. (eds.), Transboundary Water Resources in the Balkans, 41–53.
© 2000 *Kluwer Academic Publishers.*

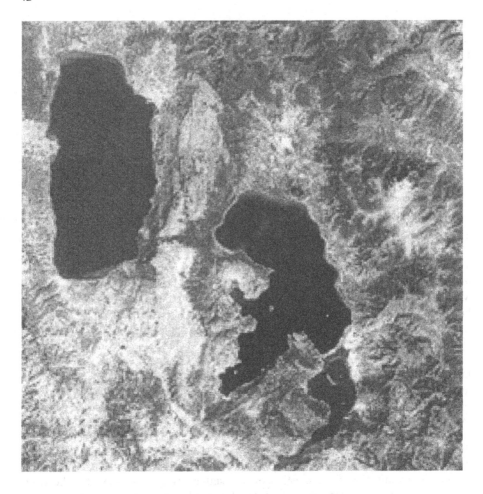

Figure 1. Lake Ohrid and Prespa (satellite view).

2.1 NATURAL CHARACTERISTICS

Approximately two thirds of the lake surface area belongs to FYROM and the remaining third to Albania. (Figure 2). It has a surface area of 358.2 km ^{2}, a maximum, depth of 288.7 m, a mean depth of 163.71 m, a water volume of 58,6 km^2, and a shoreline length of 87.53 km (56.02 km in FYROM and 31.51 km in Albania). It has a drainage area of 1129 km^2 and 40 tributaries. The most dominant of these are temporary creeks and rivers, especially during heavy rains and from snowmelt from the surrounding high mountains. Due to the karstic underground, water from Lake Prespa and its catchment area (situated to the east of Lake Ohrid behind the Galicica Mountain, and 150 m higher than Lake Ohrid) contributes significantly to the supply of several springs along the shore of Lake Ohrid. The effective catchment area of Lake Ohrid thus extends into the catchment area of Lake Prespa, and may exceed 2,000 km^2 (Figure 2).

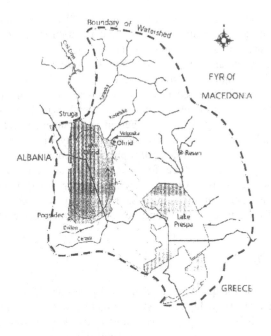

Figure 2. Main inlets and outlet of Lake Ohrid and its watershed boundary.

This lake is probably unique in that its water supply comes primarily from numerous surface and sub-lacustrine springs, mostly located on the eastern and southern parts of the lake. Surface springs mostly occur in clusters along the southern borders of the lake.

By using the isotope tracing method it can be shown that over 56% of the water originates from the neighbouring Lake Prespa, which has a maximum depth of 46 m and is separated from Lake Ohrid by the karst mountain of Galicica. Its surface is 845 m above sea level or around 150 m above the surface of Lake Ohrid.

The lake has only one outlet, the River Crni Drim (22.24 m³/sec), which belongs to the Adriatic drainage area. This outlet is manually controlled and lake water can be released up to a maximum level of −54 cm. Further downstream two dams for electricity power stations have been constructed.

2.2 PRODUCTIVITY

From the view point of productivity Lake Ohrid is a typical oligotrophic lake. According to Petrovic [1] the low values of the total phosphorus and nitrogen in the lake water are due to the poor content of nutrients in the pedological substrate of both the lacustrine basin and the drainage area, since they are situated in a karstic region. Compared to other lakes, Petrovic also found low values of total and organic phosphorous, reflecting the quantity of phosphorous and nitrogen fixed within phytoplankton and detritus. These low values are further indication of the poor production capacity of the lake, and of how very small an autochthonous supply of phosphorous and nitrogen is capable of being regenerated through bacterial activity into the inorganic nutrients of phosphates and nitrates.

According to Allen and Ocevski [2], the integrated organic productivity per m^2 water column of the trophogenic zone from the surface of the lake to a depth of 100 m has relatively small values of between 63 $mgC/m^2/day$ in the winter period, and 400 $mgC/m^2/day$ during summer months. They also found that there was a significant increase in productivity during the summer period in the layers between 60 and 100 m deep. This is probably due to the photosynthetic activity of the phytoplancters with oligothermic development and their heterothrophic nutrition ability.

Lake Ohrid represents a refuge for numerous freshwater organisms from the tertiary period, whose close relatives can only be found as fossil remains. This is why the lake is known as a "museum of living fossils." It is inhabited by numerous endemic and relic forms of organisms, which contribute significantly to the importance of the lake.

2.3 FISH RESOURCES

The size of the lake and the quality of its fish fauna give Lake Ohrid added significance The fish are of much better quality and higher value than in the rest of the Balkan lakes, even exceeding the shallow and productive lakes in the Aegean zone (Stankovic [3]). The lake has been characterised as being a typical salmon lake, and fishing is an important industry in this part of the country.

Artificial spawning of the native Lake Ohrid trout was begun in 1935. Since then, the lake has been stocked with more than 450 million young trout in different stages of development. The fish hatchery was established as part of a hydrobiological station. Traditionally the investigations of the Hydrobiological Institute (HIO) were focused primarily on measurements of nutrients, temperature (currents and mixing), transparency, primary production, chlorophyll, flora and fauna, problems of endemism, and speciation. In the past three decades research on pollution and eutrophication of the lake has resulted in the publication of over 700 papers.

3. Eutrophication of Lake Ohrid

3.1 LEVEL OF EUTROPHICATION

A main characteristic of Lake Ohrid's ecosystem is the scarcity of nutrients and consequently the low level of primary production in the lake. A vast layer of phytoplankton reveals the oligotrophic condition of Lake Ohrid. The great transparency of the lake waters - a common feature of oligotrophic waters - allows deep light penetration, which is a main factor determining the extent of the productive layer. Stankovic [3] reported that the productive layer reaches a depth of 50 or even 75 m during summer stratification; some species of phytoplankton were even found as deep as 200 m below the surface.

The eutrophication of waters (i.e. the increase of its primary production) may in principle be limited by the availability of each of the three main components of algae, i.e. organic carbon, nitrogen and phosphorus. Many scientists believe that it is the concentration of phosphorus, which controls the growth of algae in oligotrophic waters. Organic carbon and nitrogen may be limiting only if the phosphorus content is very high, such as in human wastewater or in waters which consist to a large degree of human wastewater.

It is assumed that currently phosphorus is the limiting factor of algal growth in Lake Ohrid Thus phosphorus concentration is used as an indicator of eutrophication. Monthly measurements between June 1988 and July 1989 reported by Naumovski [4] revealed a mean concentration of total phosphorus of 6.1 mg/m^3. The mean gradual annual increase in concentration of about 0.25 mg/m^3 indicated by recent studies, would result in a current concentration of between 7 and 8 mg/m^3. The thickness of the productive layer, which depends mainly on the degree of light penetration and temperature, shows that the compensation depth for some representatives of the phytoplankton in Lake Ohrid, i.e. the depth at which the algae can produce enough organic matter for phytoplankton growth, can reach 200 m. The mean chlorophyll-a concentration of 1.59 mg/m^3 measured by Mitic [5] (1990) also clearly showed the oligotrophic character of Lake Ohrid, when comparing this value to the typical range of chlorophyll-a concentrations (1 to 3 mg/m^3) reported for oligotrophic lakes. In order to assure oligotrophic conditions in Lake Ohrid in the future, the mean concentration of total phosphorus in the lake water should be maintained at below 7 mg/m^3.

3.2 SOURCES OF EUTROPHICATION

The phosphorus input into the lake comes partly from human activities (e.g. wastewater production, inadequate fertilising and cultivation of land), and partly from natural sources (i.e. expected input from an ecosystem undisturbed by human activities). To estimate the phosphorus input, the following contributions are taken into account:

- human wastewater and detergents
- Run-off, wash-out and erosion within the watershed.
- Atmospheric deposition on the lake surface.

Erosion brings particulate phosphorus, which normally is not readily available for algae growth, but may become so over a long period of time. The other sources bring phosphorus mainly in dissolved or solid form easily available for algae. The input into the lake may be point source in the case of rivers, springs or wastewater collectors discharging into the lake; or diffuse (non-point source) if the input is through non-permanent watercourses or through direct atmospheric deposition on the lake surface. It was estimated that total phosphorus input for 1995 amounted to 250 tons, stemming from non-point and point sources. This figure reveals that the wastewater of Pogradec and its surrounding villages is probably the current major point source of phosphorus. The Sateska River and the springs between St. Naum and Tushemisht are believed to be other important point sources, fed mainly by diffuse phosphorus input from their watersheds.

The phosphorus input from the River Sateska is believed to come mainly from eroded material. The estimation for the springs is based on recent measurements of the phosphorus concentration in Lake Prespa by the HIO. While these measurements in general show a lake in mesotrophic condition (mean total phosphorus concentration around 20 mg/m^{-3}), measurements during the autumn revealed significantly higher concentrations in deeper levels of the lake (concentration reaching 65 mg/m^{-3} at a depth of 18 m). This increased phosphorus content could be caused by:

(i)   an increased run-off/wash-out of fertilisers from the cultivated land around Lake Prespa and/or
(ii) by the decreased water level of the lake.

Reason (i) seems possible since the use of fertilisers in the orchards around Lake Prespa is reported to be high. Reason (ii) should be taken into consideration, since the decrease in the water-level has drained wetlands south of Resen, which may have served as a natural nutrient filter for the in-flowing treated and untreated wastewater. At the same time, the falling water level has diminished the lake's volume by about 10%, thus similarly increasing its phosphorus concentration. If the trend to anaerobic conditions in Lake Prespa continues, a release of phosphorus from sediments could become important and could increase the phosphorus content of Lake Prespa dramatically. Further investigations are needed to corroborate these hypotheses.

The total phosphorus input into Lake Ohrid was estimated for three scenarios:

- Scenario 1: population as of 1995.

- Scenario 2: population as expected for 2025; Lake Prespa as of 1995 (mesotrophic); wastewater treatment as of 1995.

- Scenario 3: population as expected for 2025; eutrophic Lake Prespa; wastewater treatment as of 1995.

Scenario (1) reflects the present situation; scenario (2) gives an estimate for the situation in 30 years time, assuming a population growth but no extension of the wastewater treatment systems; scenario (3) simulates the expected situation in 30 years time if Lake Prespa turns strongly eutrophic, i.e. with a total phosphorus concentration of 100 mg·m^{-3}.

The phosphorus input today (scenario 1) is thought to be up to 38% in particulate form (due to eroded material) and up to 62% in dissolved form. Dissolved phosphorus is estimated to stem primarily from detergents (45%) and human sewage (23%). Assuming a population growth and no improvement of the wastewater treatment, the importance of sewage and detergents as phosphorus sources will increase in the future. Other sources of dissolved and particulate phosphorus are expected to stay nearly constant (scenario 2). If Lake Prespa were to turn strongly eutrophic (scenario 3), the springs from this watershed may become a major source of phosphorus for Lake Ohrid. Phosphorus is subject to physical, chemical and biological reactions in the lake. It may take the form of organic or inorganic, dissolved or solid phosphorus, and stay in the water body or sink to the lake bottom.

4. Environmental Monitoring Programme

4.1 PROGRAMME NEEDS

A monitoring programme for Lake Ohrid should ensure early detection of conditions that necessitate particular measures, and provide information on the progress and results of any such protective measures undertaken.

The establishment of a comprehensive monitoring and controlling system is needed to enable an effective identification of problems, to evaluate the efficiency of all measures undertaken and to identify the need for future measures. These requirements can only be met with a monitoring and controlling system which:

- Allows the recording of the most sensitive indicators to show the slightest changes in water quality and the corresponding reaction of the ecosystem.
- Reliably predicts significant changes by means of long-term measurements of all appropriate parameters.
- Enables the indication of local problems by means of a monitoring network of sufficient spatial resolution.

It is expected that the Hydrobiological Institute will meet all the operational requirements of such a monitoring system on the lake and its tributaries. Some conceptual support may be needed to install the monitoring system that the Swiss Federal Institute for Environmental Science and Technology may be asked to provide. This institute is in charge of many similar tasks in Switzerland and other countries.

4.2 SENSITIVE ECOSYSTEM INDICATORS

The most sensitive parameters of an ecosystem must be included in any monitoring programme to recognise important changes and to enable the interpretation of anthropogenic impacts. The most sensitive part of the Lake Ohrid ecosystem is its biological structure and diversity of the numerous endemic species. However, assessment of the biological system is the most difficult and time-consuming part of the monitoring programme.

Being convinced that a healthy habitat provides a healthy community of plants and animals, any changes in the environment should be monitored. This is much easier than monitoring all biological aspects. The most sensitive parameters are listed in below in Table 1 below.

4.3 SUGGESTED MONITORING PROGRAMME

A prerequisite is the evaluation and interpretation of existing data. This aspect is considered as part of the monitoring programme. By comparing the results of new investigations with earlier results, such as species lists, threatened and endangered (endemic) species can be recognised, and anthropogenic impacts may be changed accordingly to avoid further harmful changes in the environment. Of primary interest are the phyto- and zooplankton, benthic fauna, fish, reed belts and *Chara* meadows in the littoral and sub littoral zone, respectively. The proposed monitoring programme consists both of permanent and sporadic measurements and investigations.

TABLE 1. Indicators and anticipated impacts.

INDICATOR	SOURCES OF IMPACTS
Bacteria (E. coli) in the littoral zone	Local wastewater pollution
Phyto and zooplankton communities	Heavy metals and organic contaminants (like anthropogenic activity and pollution in the area, insecticides, fungicides and herbicides) (mining, agriculture, industry).
Oxygen	Primary production (oxygen over saturation in the uppermost layers of the epilimnion) and the decomposition of organic matter (oxygen depletion in the lower hypolimnion). Since complete mixing in Lake Ohrid occurs only about every 7 years, oxygen, depletion in the deep hypolimnion is an indicator of appropriate or excess production, i.e. eutrophication.
Phosphorus, nitrogen, chlorophyll a	Nutrients are the basis of plankton standing crop and production and therefore key factors of eutrophication. The dominant nutrients introduced by anthropogenic activities are phosphorus and nitrogen. Algae growth in Lake Ohrid is expected to be limited mainly by phosphorus. For short periods, other factors may become limiting, such as light, temperature and possibly silica.
Temperature structure and currents	mixing and lateral transport of nutrients and contaminants, especially from the littoral into the pelagic zone.

5. References

1. Petrovic, G., 1975: Hydrochemical study of Lake Ohrid in respect to its metabolism. Reccul des travaux. Hydrobiological Institute, Ohrid, XV, 3(84): 1-70.
2. Allen, H.L. and Ocevski, B.T., 1976: Limnological studies in a large deep oligotrophic lake (Lake Ohrid, Yugoslavia). Evaluation of nutrient availability control of phytoplankton production through in situ radiobioassy procedures. Arch. Hydrobiol. 77 (1): 1-21
3. Stankovic, S., 1960: The Balkan Lake Ohrid and its living world. W. Junk, Den Haag, Monogr.biol.9, pp 357.
4. Naumovski, B.T., 1994: Trophic state of Lake Ohrid as a result of the total phosphorous content. Skopje, PMF, Institute for Biology. Master thesis.
5. Mitic, V., 1990: Qualitative and quantitative investigations of the vertical distribution of the phtosynthetic pigments in the phytoplankton of Lake Ohrid and their significance for assesment of the trophic level state of the water. Skopje, PMF, Institute for Biology. Doctoral thesis.

Appendix 1. Frequency of Monitoring and Relevant Parameters.

Permanent Sampling Stations in the Pelagic Zone

Chemical parameters

Stations 1+2:	12 samplings per year (monthly sampling)
Stations 3-8:	6 samplings per year (bimonthly samplings)
	Vertical profiles (8-10 depths):
Stations 1,3-8:	Total phosphorus, particulate phosphorus, dissolved phosphorus (P_{tot}, PP, P_{sol})
Stations 1,3-8:	Total nitrogen, nitrate, nitrite, and ammonium (N_{tot}, NO_3 and NO_2, NH_4)
Stations 1,3-8:	Particulate organic carbon, dissolved organic carbon (POC, DOC)
Stations 1-8:	Oxygen (O_2) (Winkler)
Stations 1+2:	Chlorophyll a, heavy metal and pesticides. Concentrations in suspended matter.

Physical parameters

Stations 1-8:	Monthly samplings
	Vertical profiles:
Stations 1-8:	Temperature, conductivity, Secchi depth

50

Biological parameters

Station 1:   Monthly samplings
Vertical profiles (9 depths):
Phytoplankton, primary production (C-14)
Zooplankton
Fish:
Accumulation of heavy metals and pesticides measured in 10 species of fish important for consumption. Project as proposed by HIO for trout; once every 5 years
Fisheries:
Catch statistics for all species of fish caught, once a year.

Local Pollution and Contamination

Bacteria (E.coli) concentrations near pollution areas

Monthly samples at stations: River Cerava (c), Pestani (d), River Velgoska (f), Peninsula of Lin (h), Drilon (j), Tushemisht (k), Memelisht (l), Hudenisht (m); Survey continued at stations: Pogroda/Kalista (a), Hotel Metropol Ohrid (e), Ohrid bay (g), and Pogradec (i).

Main Tributaries, Effluents of Treatment Plants, Non-Point Sources

Load study of major rivers

Sateska (inorganic silt, nutrients and pesticides)
Koselska, Velgoska and Cerava (industrial contamination)
Canal and River in Pogradec, spring Tushemisht, spring St. Naum, and outlet River Crni Drim (oxygen)
Measurements to be taken once every 14 months (may be repeated after major changes in the catchment)
Q-measurement, continuous (limnigraph). Discharge
Q_s-samplers sampling in 18-day rhythm. Samplers made by Quantum Science company (sampling depending on the amount of discharge)
Additionally sampling of all high flow events
Analysis of P_{tot}, P_{sol}, N_{tot}, NO_3 and NO_2, NH_4; POC, DOC

Effluents from wastewater treatment plants

Once every second year.
QS-samplers: sampling in 18-day rhythm
Analysis of P_{tot}, P_{sol}, N_{tot}, NO_3 and NO_2, NH_4; POC, DOC
Sludge contamination: once a year

Agriculture in Lake Ohrid watershed: portion of nutrient input based on area statistics, and population and livestock statistics.

Sediment Cores

4 transects of cores: depth profiles 0-20 cm (to be repeated every 10 years)

Pogradec (mining) - Ohrid: littoral, sublittoral, profundal
Piskopati (Cr-mining) - Pestani
Ohrid (industrial pollution) - Lin
River Sateska (silting) - Lin

Analysing cores in terms of:
Heavy metals concentrations, POC, particulate nitrogen (PN)
PP in the solid phase
DOC, NO_3, NH_4 in the pore water
Benthic fauna species

Littoral Zone

Macrophyte species list and distribution
Phragmites and Chara every 10 years

Benthic community species inventory
Every 10 years

Changes of fish spawning grounds: littoral and sublittoral
Once a year

Appendix 2. Organisation of Monitoring Responsibilities Lake Ohrid Project.

Pelagic Zone

Parameters	Institutions in Albania				Institutions in FYROM		
	HMI	IPH	BRI	CIF	HIO	PHI	HMI
Chemical parameters	X			X	X	(X)	
Physical parameters	X			X	X		
Biological parameters		X	X		X		
Fish, fisheries			X	X	X		

Local Pollution and Contamination

Parameters	Institutions in Albania				Institutions in FYROM		
	HMI	IPH	BRI	CIF	HIO	PHI	HMI
Biological p. (mainly E. coli)		X	X		X		
Chemical p. (2nd priority)	X	X		X	X	(X)	

Sediment Cores

Parameters	Institutions in Albania*				Institutions in FYROM**		
	HMI	IPH	BRI	CIF	HIO	PHI	HMI
Chemical parameters				X	X	(X)	
Biological parameters			X		X		

Abbreviations
* HMI Hydrometeorological Institute, Tirana; IPH Institute of Public Health; BRI Biological Research Institute; CIP Civil Engineering Faculty.
** HIO Hydrobiological Institute, Ohrid; PHI Public Health Institute, Skopje; HMI Hydrometeorological Institute, Skopje

Main Tributaries, Effluents of Treatment Plants, Non-Point Sources

Parameters	Institutions in Albania				Institutions in FYROM		
	HMI	IPH	BRI	CIF	HIO	TBI	TBI
Load studies of rivers:							
Physical parameters	X				X		[X]
Q-measurements	X			X	X		[X]
Chemical parameters	X	X		X	X	[X]	
Effluents from treatment pl.							
Physical parameters	X				X		
Chemical parameters	X	X		X	X	[X]	
Diffuse sources	X	X		X	X		

Littoral Zone

Parameters	Institutions in Albania				Institutions in FYROM		
	HMI	IPH	BRI	CIF	HIO	TBI	TBI
Macrophytes			X		X		
Benthic community			X		X		
Fish spawning grounds			X		X		
Heavy metals in macrophytes			X	X	X	[X]	

X Responsible institution **(X).** Supporting institution **[X]** Possible future participation
TBI To be identified

THE NERETVA RIVER: INTEGRATED COASTAL AREA AND RIVER BASIN MANAGEMENT

J. MARGETA
Faculty of Civil Engineering
University of Split
Matice Hrvatske 15
21000 Split, Croatia

1. Introduction

River basin management often extends over national boundaries and is thus sometimes subject to international conflicts. Coastal zone management is often local, regional and international. The Neretva River's basin and coastal area have created an international management problem between Bosnia-Herzegovina (B&H) and Croatia with multiple issues at stake, such as pollution of the estuary and sea, flood protection, water supply, and the management of special protection of estuary and marine areas.

The Neretva River is of great socio-economic and environmental importance for B&H and Croatia. It is a river with significant hydro-energetic potential and production, and an important source of water for supply, irrigation, and flood control. The lowland part of the river, which consists of wetlands, and the estuary, are very important protected ecological areas. Part of the coastal sea under the influence of the river is also a protected area for shellfish growing. The majority of the watershed is situated in B&H while the lowland area and estuary are in Croatia. Coastal areas affected by the river are located in both B&H and Croatia. These administrative boundaries in river basin and coastal areas make the solution of common problems, and the optimal management of the river basin and coastal areas, exceptionally complex and difficult.

The situation calls for the consideration of integrating river basin and coastal area management issues, providing both areas with opportunities for treating management needs in a more direct way than in the past, so that they may be regarded as part of a unified framework and management system. This will ensure improved co-ordination of policy making across sectors (water, forestry, agriculture, urban development, environmental protection, etc.) and on the spatial level (coastal and estuary zones to river basin) ultimately leading to a more rational use of resources and more effective environmental protection.

Achievement of an integrated concept of water resources management and the sustainable development of river basin and coastal areas depend on integrated plans for water resources management. The main prerequisite for sound planning and water resources management is sufficient data sources, i.e. an overall monitoring system. An integrated monitoring system should be established that treats all usual and special

J. Ganoulis et al. (eds.), Transboundary Water Resources in the Balkans, 55–65.

problems and characteristics of an integrated system river basin-estuary-coastal area in a co-ordinated way.

This paper presents the problems associated with the integrated management of the Neretva River and its coastal strips; and provides guidelines for a co-ordinated approach that will support the development of a transboundary monitoring network.

2. Main Characteristics of the Area

The Neretva River can be divided into two main parts, the upland part of the river basin and the estuary. The coastal sea area is strongly influenced by the river, which is the largest in B&H and Croatia belonging to the Adriatic Sea basin. It is also one of the most important fresh and coastal water resources and ecological areas of international significance on the Adriatic coast. The lowland part of the Neretva River is an important route to and from the sea, and from the sea B&H has access to the rest of the world.

2.1 THE RIVER

The Neretva River basin is predominantly a karstic area, and it is thus very difficult to determine exact watershed boundaries. The length of the river is 240 km with a catchment area of about 10,100 km^2 (Figure 1). The river has numerous tributaries, which influence the main stream either directly on the surface or indirectly in ground water. The mean level of the river basin is about 250 m above sea level (m.a.s.) with the highest point of the river at about 900 m.a.s.

River water flows mainly through limestone in accordance with the hydro-geological characteristics of the karst flow. As a result of the pronounced karstic phenomena vegetation is poor in the river basin and there are few cultivated areas. Significant lowland areas are karstic poljas (openings) on different altitudes that are hydrologically interrelated. The majority of poljas can be cultivated, although floods endanger agriculture during the winter months.

Climate characteristics in the river basin vary with the distance from the sea. The climate is Mediterranean in the lowland area close to the sea, while the middle part has a continental climate and higher areas have a mountainous climate. The average rainfall is 1,650 l/m^2, and varies between 1,500 and 1800 l/m^2. Most of the precipitation falls in winter, in November, while a minimum occurs in summer, in July, which is often without precipitation. Temperatures range between -29 and +43 °C, and the annual evapo-transpiration is 500-900 mm.

The average discharge is 269 m^3/s, while minimum and maximum discharges are 44 m^3/s (probability 0.05) and about 2,179 m^3/s (probability 0.01), respectively. The runoff coefficient is about 0.871. Discharge has been measured at 21 gauging stations for more than 30 years. Water quality is measured on several sections of the river and on significant tributaries. More than 323,000 inhabitants live in the area of the river basin, and the industrial pollution load is estimated to be 250,000 ES. The water quality is satisfactory except for the sections immediately following large settlements.

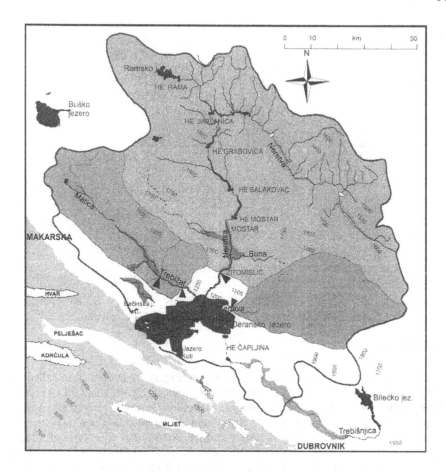

Figure 1. The Neretva River basin.

The most important role of the river is hydro-energy production. There are five hydroelectric power plants in the river basin, and seven more are planned (Figure 2). The water storage reservoirs adjacent to hydroelectric power plants are multipurpose. After hydro-energy production, their next most important role is flood protection. The river has beautiful surroundings and has good water quality, especially in the upland areas, and thus represents an important resource for development of the wider region.

However, water accumulation also has adverse environmental impacts especially on the estuary and coastal sea. Unfortunately, the majority of towns and industrial plants still do not have wastewater treatment plants. The lack of treatment has a significant impact on the lowland areas. Such pollution includes wastewater from chemical plants and the metallurgical industry in the area of the city of Mostar, which is the most important industrial centre in the river basin. Pollution affects the use of river water for supply and irrigation. Because of the karstic features of the area, water is not greatly purified by underground flow, and the river is highly susceptible to pollution. It should be stressed that the majority of the river and its basin are situated in B&H.

58

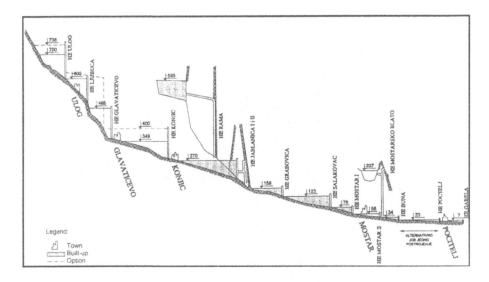

Figure 2. Hydroelectric power plants (HPP) on the Neretva River.

2.2 ESTUARY

The major part of the estuary is situated in Croatia (Figure 3) and has been called the "Delta of the Neretva River". The estuary covers an area of 156 km^2, and the river is 19 km long in this area. In the estuary the river branches into several rivulets and channels, which flow into the sea, forming the estuary system. In the boundary area there are also several significant springs that recharge surface water and thus form a unique hydrologic system for the estuary. The entire area is relatively low from −1.0 m above sea level (a.s.l.) to 6 m a.s.l. In the past, the majority of the area was regularly flooded in winter.

In the last 50 years the area has been improved and protected against floods. Flood protection levees were constructed and protected areas are used for intensive agricultural production. A mild Mediterranean climate, fertile soil and water abundance make this area especially favourable for agricultural production throughout the year. It is the area with the highest production of citrus fruits in Croatia, as well as having winter and early spring crops. Most of the agricultural production is exported to B&H.

None of the area has been drained and so significant areas of natural marshes have been retained as natural ecological wildlife refuges. The area is ecologically very significant, and together with the upland marsh area in Bosnia-Herzegovina, forms one of the most important coastal marsh areas in the Mediterranean. The area is rich in fish, birds, and other wetland flora and fauna species.

The plain has more than 65,000 inhabitants living in several settlements, the largest being Metkovic and Ploce. The main economic activities are agriculture and transport services. Ploce is an important harbour on the Adriatic coast and the main port for the whole of B&H. The river is navigable as far as Metkovic, which is situated on the border with B&H; and is therefore also a river harbour with a significant infrastructure of railway, roads and an airport. This area forms the crossroads between the Adriatic coast and the main route from the coast to inland B&H. There is no large industry except some in the Ploce harbour and no significant industrial pollutants. Pollution is the result of municipal wastewater and the drainage from roads and agricultural development. The most significant pollution posing the greatest danger to the ecosystem comes from the upland areas of Bosnia-Herzegovina. The quality of water in the river and its tributaries has been continuously measured for more than 20 years. These records provide a large database.

2.3 COASTAL SEA AREA

The estuary is part of the coastal area of the Adriatic Sea called the "Neretva Channel". Although called a "channel", it is actually a deep and narrow bay, only 6 km wide at the mouth, and 30 km from the mouth to the bay end. The Neretva flows into the sea, and provides a slow sea water exchange. At the end, the bay is one of the most unpolluted and well-protected parts of the Adriatic Sea, and is used for shellfish growing especially oysters. It is an area where the construction of tourist facilities is forbidden, which is not the case in the area closer to the Neretva River mouth.

Most of the area under the influence of the Neretva River belongs to the Republic of Croatia; the smaller part belongs to B&H. In fact, the entire coastal area of B&H is part of that area (Figure 3). The only coastal settlement of B&H is Neum, which is the largest settlement of the area except for the Ploce harbour, situated to the west in the immediate vicinity of the Neretva River mouth. There are several smaller tourist settlements in the coastal area to the south and west of the mouth, as well as on the other side of the bay on the Pelješac peninsula. Besides tourism, other important activities are fishery and aquaculture i.e. shellfish growing. There is also the aforementioned Ploce harbour. The main Adriatic coastal road also extends along this area in the direction of Dubrovnik and Greece.

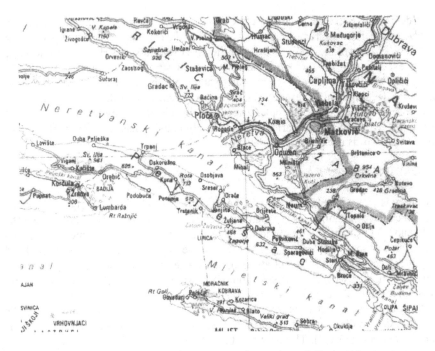

Figure 3. Area of the Neretva River estuary and coastal part influenced by the river.

2.4 IMPACTS AND LINKAGES

The river basin, estuary and associated coastal sea form a unique hydrological system within which a series of natural and socio-economic processes occur. The river basin provides an inflow of fresh water, nutrients and sediment to the estuary and the sea, but also pollution from upstream industry, settlements and agricultural activities. Development and water accumulation in the river basin cause a series of disturbances (sediment retention, change in the discharge and flooding) in the natural characteristics of estuary and coastal sea, but also provide flood protection. The coastal area is a consumer of river basin resources, and plays an important role in the flow of goods and people as well as in tourist and recreational activities of river basin inhabitants. The coastal area, which extends on both sides of the Neretva River, is the traditional Bosnian "Riviera." Tourist resources developed in this area belong predominantly to companies from B&H, and there are numerous summer homes.

The estuary, situated between the river basin and the coastal area, has a significant natural and socio-economic function. It is the most fertile area thanks to the water from the river basin, producing food crops for people living in the river basin and coastal areas as well as for other areas in Croatia and B&H.

All of these factors support the need to use an integrated approach to the management of the river basin, estuary and coastal area. Achieving such an approach is a special challenge, because administrative boundaries are interrelated through key segments of the natural system. Resources of the river basin (river, estuary, and coast) provide a flow of goods and services, as well as a number of complementary and conflicting activities. If left alone, social and economic factors at work in basin areas competing for resources would result in overexploitation of resources, negative environmental effects, equity problems and loss of social well being. A prerequisite for the sustainable management of the river basin is the development of an integrated development plan that takes into account the use of water resources, and above all includes a transboundary monitoring network adjusted to the specific characteristics of natural and socio-economic systems.

3. Monitoring Issues

Monitoring has been carried out in both the Croatian and Bosnian parts of the Neretva river basin for many years, and until the countries became independent in 1991 was guided by the respective laws of each republic in the Federation, each of which had its own monitoring and supervision system. Coastal water monitoring, however, was carried out and supervised exclusively by Croatian institutions.

The formation of the independent states created a completely new situation, where international regulations and conventions replace agreement between the republics of a single state as a basis for international river basin management. Monitoring of water systems will be carried out based on an agreement similar to that now guiding transboundary monitoring activities with Slovenia and Hungary. The situation concerning the Neretva River is more complex, since the river and the corresponding coastal sea belong to both Croatia and B&H.

The use of the water resources of the river Neretva should be divided equitably between the two countries sharing the Neretva catchment and associated coastal waters. Protection of the quality of water resources need to be guaranteed and their sustainable use maintained through an effective mechanism. Bi-lateral transboundary agreement and regulations always reflect compromise by the two countries involved.

3.1 AN INTEGRATED PLAN

An integrated plan is needed for the sustainable management of the river basin, including river, estuary and coastal sea. It should create a technical, administrative and political basis for the establishment of a monitoring system as well as appropriate co-ordination mechanisms

Since both Croatia and B&H are neighbours of European Union (EU) countries, they should use and implement the regulations, practices and norms of the EU. Croatia has already adjusted the majority of its legislation dealing with water and sea quality to the standards of the EU The following conventions should also provide a basis for bilateral agreements:

- Convention on the Protection and Use of Transboundary Watercourses and International Lakes (Helsinki, 1992).
- Convention on the Transboundary Effects of Industrial Accidents (Helsinki, 1992).

62

Since both countries own a share of the sea belonging to the Mediterranean basin, the agreements should include all the directives and protocols signed and accepted by Mediterranean countries, such as the Barcelona Convention and related protocols.

In order to prepare for a satisfactory international agreement, Croatia has already started certain preliminary actions such as an assessment of resources of the river Neretva estuary and coastal areas, which was completed in 1996 by the Faculty of Civil Engineering in Split. Conclusions and recommendations were drawn up, among which are the following:

- Influence of upstream sections of the river is dominant and crucial for the state of the estuary and coastal sea, therefore, the river basin has to be reviewed in an integral way and agreement reached on the sustainable use of its resources
- The basis for short- and long-term problem solving is the establishment of an appropriate monitoring system and the co-ordination of the use and management of river basin resources.

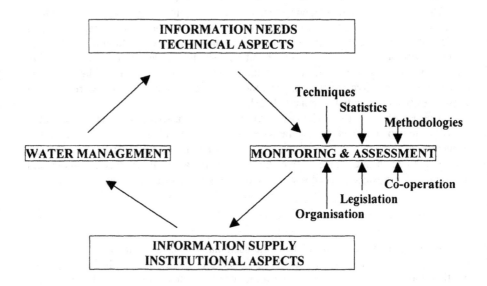

Figure 4. Tools for monitoring [1].

Guided by the experience of the past and future projections, an appropriate monitoring system is one of the most important tools in water resources management, if problems are to be permanently solved. The process of monitoring and assessment should be seen as a sequence of related activities that begins with the definition of information needs and ends with the use of the information products, Figure 4.

3.2 MONITORING AND INTEGRATED MANAGEMENT

Water management of the Neretva river basin needs to combine research, policy and operational activities with the goal of developing functions and potential uses for a water system. A water system includes the water, banks, underwater beds, relevant ground water, estuary, coastal sea, existing ecosystems and technical infrastructure. Integrated water management (IWM) therefore includes the surface water, ground water, estuary and coastal waters systems. It integrates all the various aspects of the water system: water quality, water quantity-hydrology, oceanography, environment, and the sociological, economic, financial and institutional aspects of water policy. River basin systems do not have strict and permanent boundaries. They are limited in space and time by morphological, ecological and especially functional characteristics. Within the system there are specific goals and problems related to the specific characteristics of river, estuary and coastal waters. Integrated water management also means managing the water systems coherently, in one or more control systems (supply, pollution, flood, navigation, salinity, hydroelectric, etc.), by gearing the different policy and management levels to one another as well as possible. A policy of IWM should be used for the Neretva water system.

Figure 4 shows a general information system for water management. Information can be produced in different ways. Monitoring is only one of the management tools available, but is the most important. Information can also be obtained from a combination of surveys, expert judgement, models, decision support systems and historical databases. Monitoring water systems provides systematic, permanently available information about the performance of a water system with respect to its social and ecological functions.

Any revised monitoring system of the Neretva River should build on the present system, which needs to be redefined in accordance with the integrated plan of water resources management. Periodic evaluations of the new monitoring and information system should take place. There can be problems with respect to developing such a system. The four parts of the system (Figure 5) represent four separate worlds of expertise, which often do not co-operate effectively.

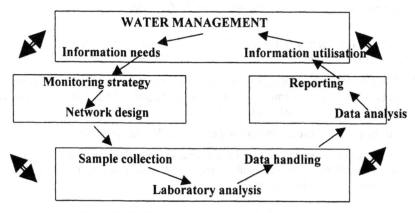

Figure 5. The four parts of the monitoring process.

Information may be passed from one to the other without genuine interest or insight into what use the data is to be put This will make integrating the Neretva's river, estuary and coastal sea monitoring systems into a single system particularly difficult. It should be stressed that such a monitoring system has never been implemented in Croatia or B&H.

4. Conclusions

Monitoring of the Neretva water system must be viewed as part of a complex management system involving water system goals and targets, water managers, public interest, commercial interests, tourism, navigation, natural conservation interest and so forth.

Monitoring systems in rivers, estuaries and affected coastal waters are seldom integrated into one system. Usually only one or two aspects of this water system type are taken into account (single-purpose monitoring). In order to establish a more complex system, good will and the cooperation of interested parties from both countries is required.

The need to integrate monitoring in river, estuary and coastal sea depends on how far processes and variables among river basin elements are interrelated. At present, without the necessary analyses and water resources plans, it is hard to say how extensive a system is necessary. Maybe it would be sufficient to just gear the information needs for all three systems to one another and to integrate the assessment. Physical integration would be very difficult and is not recommended.

When examining issues regarding the transport of water and solids/solutes, it can be useful to jointly assess information from more than one system type (ecosystem, agriculture, urban water, industry, etc.). This is especially true if the issues are nutrient transport, salinisation, eutrophication, macro- and micro-pollution, sediment transport, etc.

An integrated monitoring system must take into account all similarities and differences between the three basic elements of the river basin, namely river waters, estuary and coastal seas. The similarities are the following:

- River water, estuary water and affected coastal seas are parts of the same hydrological cycle.

- Often all three monitoring systems deal only with a part of water management and are not integrated. Within the framework of integrated water management there is need for more monitoring integration.

- All systems can use the monitoring cycle to tailor the monitoring systems.

- River water, estuary and coastal sea have many functions (ecological, socio-economical) and many issues in common (pollution, ecosystem preservation, sediment and nutrient control).

Differences between monitoring river waters, estuary and coastal seas are the following:

- Physical characteristics of the subsystems are quite different: flow velocities and reaction time in rivers, estuaries and the sea are different.

- Salinity of water is different from salinity of seawater, transient salinity in estuaries and no salinity in river.

- Water systems with the same functions can have considerably different measuring devices, layout of measuring locations, the same measuring variables and monitoring frequencies.

The preceding analysis supports the recommendation that a water resources plan for the Neretva must precede the creation of a regional monitoring network. Once accomplished, the countries should continue to collaborate on the implementation of the plan and on developing the structure of the monitoring network for river, estuary and associated coastal waters.

5. References

1. Adriaanse, M. (1997) Tailor-made Guidelines: A Contradiction in Terms, European, 7, No. 4, July 1997.
2. Claessen, F.A.M (1997) Comparing monitoring of surface and ground water systems, European, 7, No. 4, July 1997.
3. Coccossis H., Burt T., Van der Weyde J. (1999) Integrated Coastal Zone and River Basin Management (draft), UNEP - MAP/ PAP; Split, 1999.
4. Faculty of Civil Engineering (1996) Water resources assessment of river Neretve estuary, Split. (in Croatian)
5. Hock, B. (1997) Transboundary river quality problems in Hungary, European Water Management, 1, No.5.
6. Margeta, J., Iacovides I. and Azzopardi E. (1997), Integrated Approach to Development, Management and Use of Water Resources, UNEP-PAP/RAC Split, pp vi+154.
7. Salomons, W. and Turner K. (1998). The river basin dimension of coastal zone research, LOICZ Newsletter, No.9.
8. Shultz, G.A.S. (1998) A change of paradigm in water sciences at the turn of the century, Water International, 23, No. 1.

COASTAL ZONE MANAGEMENT APPLICATIONS IN TURKEY

A. SAMSUNLU, A. TANIK, D. MAKTAV, L. AKCA
ITU, Istanbul Technical University
Faculty of Civil Engineering
80626, Maslak, Istanbul, Turkey

O. USLU
9 Eylul University,
Institute of Marine Sciences
Urla- Izmir, Turkey

1. Introduction

Coastal zone management and protection are of utmost importance for countries with coastlines, primarily due to their significant tourism activities. Coastal zones are threatened by pollution from two types of sources, point and non-point. Point source pollution is from domestic and industrial waste, with the former producing a much greater volume during the summer tourist high season. Non-point sources come from agricultural activities, marine transportation, atmospheric deposition, urban run-off, and so forth. Coastal areas are important - 90 % of all the countries in the world are coastal countries with 40 % of the world's population living there. If one defines a coastal zone as a 50 m wide strip of coastline, 50 % of the world's population is settled there [1, 2].

Turkey is surrounded by coastal seas on three sides and has approximately 8,300 km of coastline, which is among the longer coastlines in Europe. Half of its population of 65 million lives in the coastal cities, districts and villages, although the coastal area covers only 29 % of the total surface area of the country. These coastal areas are of great national and international significance and attraction, and it is extremely important that their use be carefully planned and measures taken to protect them from deterioration.

The Aegean and the Mediterranean coasts experience large numbers of domestic and foreign tourists during the six-months' summer season. The rapid growth in tourism has led to a number of environmental problems. The fundamental cause of coastal pollution in the country is an insufficient and improper treatment infrastructure, together with unplanned urbanisation. Coastal development is mainly concentrated from the Canakkale-Balikesir provincial boundary in the north along the Aegean Sea to the Antalya-Mersin provincial boundary in the south along the Mediterranean Sea. Data show that of all certified beds, 27% are in the Aegean region, 25% in the Mediterranean region, and 21% in the Marmara region. Investments in summer (vacation) houses are also concentrated in these same areas, [3]. In the coastal municipalities of Balikesir,

J. Ganoulis et al. (eds.), Transboundary Water Resources in the Balkans, 67–80.
© 2000 *Kluwer Academic Publishers.*

Izmir, Aydin, Mugla and Antalya, 32% of land is devoted to vacation houses and 14% has been designated as touristic areas.

The most significant environmental problems in these heavily impacted coastal areas are an improper and insufficient infrastructure for protecting the drinking water supply and for wastewater collection and disposal. For example, only 60% of tourist settlements receive adequate drinking water, 76% have no sewerage and another 13 % have limited sanitation. Only 5% of these settlements are satisfactorily served with both drinking water and sewage facilities. According to the 1990 census, population density in coastal provinces averaged around 127 people/ km^2 compared to 73 people/ km^2 nationally [4]. An estimated 70%- 80% of all industrial output is also generated in these coastal provinces. Pollution loads have further accelerated through the construction of marinas along the coasts. From time to time, the faecal coliform counts appear to be above the limits in Akcay, Cesme and Kusadasi. 69% of land not used for tourism is fertile and mostly of the highest quality (Class I and II). Simply put, coastal activities have started to destroy the natural landscape and create aesthetic and other kinds of pollution in many areas. Therefore, in order to control the conflicting requirements of use, protection, and enhancement in a balanced fashion, coastal management plans are essential.

Studies have been conducted on coastal zone management since the 1980s and their results support the subject matter of this paper. Four additional case studies are introduced. The first is a marine discharge system that has been in operation since 1990 at Marmaris. The second, Cirali-Belek, is a general study of coastal zone management and tourism. The third is a case study concerning economic activity that presents a new approach through airborne video-graphs of the Aegean Region of Turkey. The last is an international pilot project on the development of a computer-based coastal information system for the Turkish Mediterranean coasts, which integrates satellite and ground data.

2. Management of Turkey's Coastal Zone

2.1 INITIAL PROJECTS

The need for sewage collection and treatment facilities was recognised prior to 1980. The Bank of Provinces (Iller Bank) proposed that the most appropriate and economic treatment could be provided by the construction of sewerage systems followed by marine discharge systems. In order to install such systems, especially in coastal areas, State Universities with relevant backgrounds and expertise in the subject area were assigned the feasibility surveys and design of these systems. One of the case studies mentioned in this paper, the Marmaris example, describes one of the completed projects designed by a State University. The first attempt in coastal management in Turkey dates back to the 1980s. During the 1980s the urban population began to increase while rural population showed a tendency to decrease. In particular, migration from the eastern part of the country towards the western and southern coastal areas brought about many environmental problems. Government authorities responded with environmental protection programs.

The first two important projects were begun along the Mediterranean coast of the country. Kemer and its surroundings in the southern part of Turkey has been designated a 'Special Tourist Region' and the nearby Side region has been selected as a 'Special Tourism Planning Area.' The World Bank supports both of these separate projects, and they are supervised by The Ministry of Tourism of the Republic of Turkey The main targets of the projects are to deal with sewage collection systems and the construction of wastewater treatment plants. Another project, begun at the same time in the Cukurova region, is sponsored by the Bank of Provinces (Iller Bank), and investigates in detail the environmental problems of the region extending from Mersin to Iskenderun

One of the most important projects realised in Turkey on coastal zone management is the Southern Anatolia Environmental Project (GAC) of the Ministry of Tourism, which is supported by the World Bank. In this project, the fundamental environmental problems of the coastline from south of Canakkale to the east of Alanya, covering a distance of approximately 2,000 km, were investigated with respect to tourism, agricultural activities, sauce-demographic structure, hydrogeologic structure, receiving water quality, natural beauty, technical infrastructure, current legislation and laws, and financial conditions. Alternative solutions were offered and recommendations made using a master plan approach.

The region from Mersin to Iskenderun was previously investigated within the framework of the Cukurova Project. According to the final report of the GAC project, only 9% of the total population in the project area lived in areas with sewers, 13 municipalities had old and/or new sewage collection systems and only three of these municipalities had wastewater treatment plants[5]. Wastewater treatment plants were under construction in 15 municipalities. The rest of the municipalities had no such ongoing projects. Only in five districts were faecal coliform counts over the accepted limits for swimming and water sports.

2.2 MEDITERRANEAN-AEGEAN PROJECTS

A second important project in coastal zone management in Turkey was planned in the 1990s under the name 'ATAK' – Mediterranean- Aegean Regions Tourism Infrastructure Project for Coastal Management. The project was to deal with drinking water supply, sewage collection systems, wastewater treatment and disposal alternatives, solid waste collection, recycling and disposal, design and construction facilities, and financial and administrative studies. It was planned that these studies should be conducted in 25 watershed areas covering approximately 100 residential areas, as shown in Figure 1. However, the main idea was to overcome present insufficiencies in wastewater removal, and thus to remove health risks in coastal areas of touristic significance and supply blue flags to almost all beaches and marinas.

The infrastructure requirements of all the selected residential areas were determined in stages for 1988, 2005 and 2020. Ten watershed areas were chosen as high priority. Based on the findings of the feasibility report dated October 1992, the first three high priority areas were Marmaris, Kusadası/Davutlar and Alanya regions. Within the framework of this project the Institute of Marine Sciences and Technology of the Dokuz Eylul University in Izmir conducted detailed marine surveys to form a data bank for modelling studies.

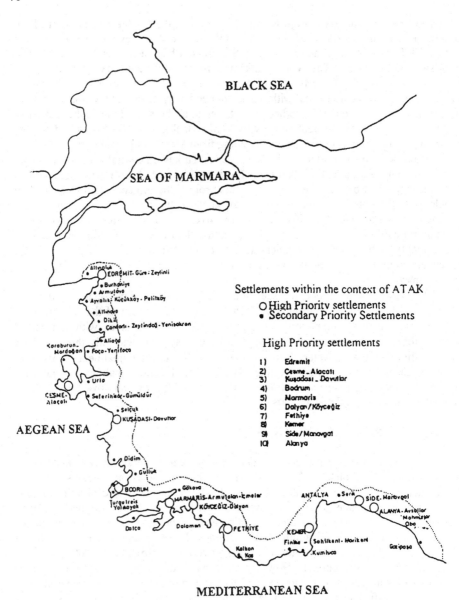

Figure 1. Settlements within the context of ATAK Projekt

The other high priority areas selected were Edremit, Cesme, Bodrum, Fethiye, Kemer, and Side. However, Kemer and Side regions had already solved about 90% of their infrastructure problems, since they were part of the 1980s programme.

Turkey's second most important coastal metropolitan area following Istanbul is Izmir on the Aegean coast. The major industries of Turkey's Aegean region are located near this city. The total pollution load to the Izmir Bay from a population of about 10 million is given in detail in the country report [6]. Many projects have been trying to save Izmir Bay for years. One such project was approved in 1981 and construction work started in 1983. At the time, activated sludge treatment was the treatment technology applied to this area. The treated effluent was supposed to be finally discharged to the Gediz area for irrigation purposes. However, due to the high costs involved, the Mayor of Izmir authorised some changes to the original treatment scheme of full treatment. Staged lagoons were then attempted. Unfortunately, this project was not approved, as it was not found to be feasible.

There are nearly 3,000 municipalities in the whole country of which 16 are Metropolitan Municipalities [5]. From 1980 on, changes in laws have made the municipalities financially stronger and more powerful. Thus, many of the municipalities, especially the Greater Metropolitan, have started to solve their wastewater treatment and disposal problems beyond water supply activities. Apart from the municipalities, other related authorities like the Ministry of the Environment (established in 1991) and its related department of Special Environmental Protection, the Ministry of Public Affairs and its related authority the Bank of Provinces, the State Water Works Administration under the Ministry of Energy, the Ministry of Industry and Commerce, the Ministry of Tourism, and the Ministry of Agriculture and Village Affairs all supply technical and financial support to realise the construction, operation and control of the so-called systems.

2.3 PRESENT STATUS OF THE AEGEAN AND MEDITERRANEAN COASTS

The following data were obtained from the Bank of Provinces in 1998 to summarise the previous efforts at improving the situation in the regions under the ATAK project. The total population of the residential areas within the ATAK project is 8,291,143, of which 28% had modern sewage systems. Twelve of the coastal districts have marine discharge systems. Only one has a wastewater treatment plant. Due to financial constraints, the districts that have completed their sewage collection systems prefer deep-sea outfall design for final removal of their wastewater. Bodrum's treatment system is nearing completion, Burhaniye's project is finished and the projects of Edremit, Anamur, Alacati, Cesme, Urla, Datca and Turgutreis are on the way, [7]. Aliaga Refinery and other important industries have their own individual treatment plants along the Aegean coast. Mersin Atac Refinery, Iskenderun Iron and Steel Manufacturing Plants and other industries along the Mersin and Adana coast in the Mediterranean region also have their own wastewater treatment facilities. The present situation of wastewater treatment plants along Turkey's Aegean and Mediterranean coasts [7] is shown in Tables 1 and 2 respectively.

TABLE 1. Wastewater treatment and discharge applications
along the Aegean coast

	Biological Number	Population	Sea outfall Number	Population
In service	8	727 084	13	301 249
Under construction	6	605 575	6	118 737

TABLE 2. Wastewater treatment and discharge applications
along the Mediterranean coast

	Biological Number	Population	Sea outfall Number	Population
In service	3	118 976	2	112 787
Under construction	1	58 104	1	46 295

2.4 THE PROBLEM OF SUMMER HOUSES (SECONDARY VACATION HOMES)

Owners of summerhouses, holiday resorts and other tourist hotels in the districts without sewers are urged to solve their wastewater problems. Some of them have constructed and operate individual treatment plants, while others discharge their wastewater into receiving water without treating it at all. However the majority prefer to collect their wastewater in simple septic tanks.

There are about 3,000 such clusters of summer settlements of which approximately 30% have their own wastewater treatment plants. They are generally either extended aeration activated sludge biological treatment plants, or low rate trickling filter systems, biodisc or sea outfall discharge systems using mechanical treatment systems. They are usually constructed as package treatment systems. However, even though these settlements are quite near each other, they have constructed individual plants instead of adopting common plant solutions. Having so many individual plants leads to great losses in investment and operational costs. Furthermore, many of these individual plants are unable to operate properly, and as it is difficult to monitor and control them, their pollution effects on the receiving water environment are still unknown. Common treatment application and sea outfall application should be adopted to lead to better monitoring and control.

Despite these positive developments, environmental issues have not been adequately incorporated into economic and social decisions. The National Environmental Action Plan (NEAP) was prepared in 1997-1998. It responds to the need for a strategy and can supplement the existing Development Plan with firm actions to integrate environmental protection and development [3]. The NEAP should be implemented over a 20-year period and comprises only short (5-year) and medium term (10-year).

3. Coastal Zone Management Case Studies

3.1 MARMARIS WASTEWATER TREATMENT PLANT AND DISCHARGE SYSTEM

Marmaris is located on the south Aegean coast of the country and is one of the most important tourist centres. Table 3 shows the projected population of Marmaris according to the ATAK Project.

TABLE 3. Projected population of Marmaris according to ATAK Project

Fixed population	Year	Population of secondary houses	Tourist bed capacity	Tourism service population	Total summer population
29 378	1990	3 524	40 864	1 308	75 073
45 160	1998	4 065	62 593	-	111 818
61 919	2005	4 065	88 970	-	139 771
80 773	2020	4 065	103 490	-	193 295

Until the 1980s the district did not have sewers. The wastewater was either directly discharged into the sea or collected in septic tanks. It was at this point that the Bank of Provinces began the construction of sewage collection systems where separate systems are preferred. The main collectors were constructed parallel to the seashore with five pump stations to aid collection. The wastewater is discharged into the sea by gravity. The sea outfall design work started in 1982 by the Civil Engineering Faculty of the Dokuz Eylul University. The treatment plant scheme prior to sea outfall consisted of screen and grit chamber units and is shown in Figure 2. It has been proposed to extend the system in the future through biological treatment. The sewage collection system and sea outfall systems have been completed and are in operation [7,8]. The discharge pipeline is made of glass reinforced plastic with a length of 740 m, a diameter of 600 mm. and a diffuser depth of 36.50 m., as shown in Figure 3. The project started in 1985 and construction was completed in 1990. No progress has been achieved on the construction of the biological treatment plant to date. However, eutrophication problems are easily detected at the discharge point and within a diameter of 200 m. Studies performed in the area show that the application of the proposed biological treatment plant would be sufficient to protect the receiving water.

The water quality monitoring studies at Marmaris Bay, following the operation of the sea outfall design, were accomplished at 33 sampling points that tested various parameters such as ammonia, nitrite, nitrate, total nitrogen, total phosphorous, SiO_3, BOD_5, chlorophyll-a, temperature, electrical conductivity, turbidity, pH, dissolved oxygen, total coliform and faecal coliform during the winter, spring and summer seasons of 1993 [8]. Table 4 shows a high total coliform concentration at some locations during the winter of 1993. The sampling points for 1993 monitoring were very close to the shore. Considering the BOD_5 and DO concentrations, it can be concluded that the water quality has a slight tendency towards deterioration. These limited data also show that chlorophyll-a concentration, as a parameter of primary production, is phosphorous limited. The N:P ratios indicate that primary production is limited by nitrogen and that there is a relative excess of phosphorus indicating that the shoreline is affected by land based sources of pollutants.

It can be concluded that the minor pollution occurring around the discharge area could be prevented by the construction and operation of the second stage biological treatment plant before marine outfall application.

TABLE 4. Water quality in Marmaris Bay 1993 [8]

Parameters	Winter. 1993.			Spring. 1993			Summer 1993		
	Av.	Min.	Max.	Av.	Min.	Max.	Av.	Min.	Max.
NH_4 (µmol/lt.)	0.67	0.00	2.79	0.36	0.00	0.95	0.48	0.10	2.43
NO_2 (µmol/lt.)	0.12	0.00	1.07	0.04	0.00	0.33	0.07	0.01	0.19
NO_3 (µmol/lt.)	1.42	0.13	27	1.28	0.00	8.71	1.51	0.01	3.43
T.N. (µmol/lt.)	2.21	0.32	30.57	1.67	0.19	9.97	2.05	0.27	4.56
PO_4-P. (µmol/lt.)	0.23	0.00	1.09	0.26	0.00	0.81	0.21	0.01	0.84
SiO_3 (µ g/lt.)	619	28	6094	392	59	1685	151	11	1152
BOD_5 (mg/lt.)	1.12	0.50	2.80	1.09	0.40	2.71	0.72	0.02	2.80
Chll-a (µ g/lt.)	0.06	0.00	0.14	0.41	0.06	3.07	0.34	0.06	0.90
EC (µmho/cm)	41.86	12.20	48.10	60.84	47.2	63.3	53.26	49.9	53.9
Turbidity (NTU)	6.34	2.60	25.00	13.39	4.00	30.0	2.47	0.50	9.00
pH	8.27	8.04	8.43	8.35	8.20	8.47	8.1	7.96	8.40
DO (mg/lt.)	8.54	7.92	9.76	7.06	6.28	8.23	6.23	5.52	7.94
T. Coli.(1/100ml)	180	0	4400	158	0	1060	77	0	550
Fec.Coli. (1/100ml)	39	0	900	48	0	710	34	0	470

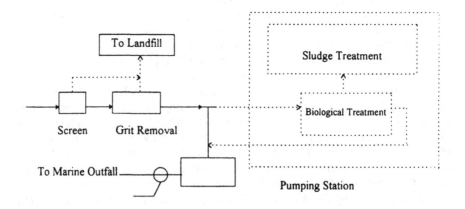

Figure 2. Flow diagram of the wastewater treatment plant prior to marine discharge system

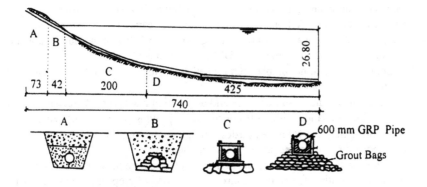

Figure 3. Schematic longitudinal and cross-sections of the pipeline

3.2 PROJECT ON COASTAL ZONE MANAGEMENT AND TOURISM

The Cirali / Belek / LIFE TCY96 /TR/ 021 project was started in February 1997 and ends in the year 2000. The European Union (EU) covered the majority (93%) of the total costs and the World's Wildlife Foundation (WWF) and the National Association of Natural Life Protection (DHKD) contributed the rest. The Ministry of Tourism, the Ministry of the Environment and the Ministry of Culture of the Turkish Republic also supported the project by providing specialist help. The project goals were to:

- Protect the coastal ecosystem including flora and fauna.
- Educate inhabitants on the importance of environmental protection.
- Gain public participation in matters regarding the protection of the regional socio-economic structure.
- Demonstrate environmental balance with the installation of a sustainable development model.

The pilot areas of interest are Antalya / Belek and Olimpos / Cirali. International tourism activities are well established in Belek, whereas in Cirali such activities have only recently started up and are being developed by the local inhabitants. This project will therefore be able to make a comparable environmental evaluation possible. The project is being run by a multidisciplinary team of scientists and experts, including urban planners, economists, sociologists, biologists, lawyers, agricultural engineers, environmental engineers, and computer scientists.

The basic areas of activity covered by the framework of the project are [9]:

- Developing and applying feasibility plans for environmental protection.
- Physical planning for Cirali.
- Management planning for Cirali.
- Management planning for Belek.

- Management planning for two reserve areas, Kumkoy and Tasliburun.
- Preparing a Priority Action Plan for the next five years.
- Studies on the applicability of current legislation and laws of the Environmental Action Plan based on a Protection-Utilisation balance approach.
- Maintaining co-ordination between related authorities and establishing regional monitoring associations.
- Determining natural resource areas and developing management mechanisms for sustainable utilisation within the concept of improving protection activities,
- Monitoring the survival of sea turtles through continuous record-keeping and planned control of any negative effects on their lives.
- Developing sustainable economic activities such as organic agriculture, eco-tourism, utilisation of natural resources, cultivation of crops important to any one region etc.
- Forming environmental socio-economic and ecological monitoring programmes.

So far, technical studies on layout planning for protection purposes, research on ecological background, administrative and organisational structure, socio-economic surveys, and environmental structures have been completed for Cirali together with preliminary surveys on organic agriculture and eco-tourism.

3.3 ECONOMICAL USE OF AIRBORNE VIDEO-GRAPHS: THE AEGEAN REGION OF TURKEY

An aerial survey to monitor the global changes in Greater Metropolitan Izmir and its surrounding regions of about 11 000 km^2 on a scale of 1/ 25 000 was requested by the Municipality of Izmir, which is situated on Turkey's Aegean coast. This case study was conducted by a group of scientists from the Institute of Marine Sciences and Technology in Izmir. The basic aim of the study was to plan an economical way to undertake air surveys in large regions. The criteria used to select an appropriate digital video-camera included digital image transfer capability and CCD structure of the camera, the pixel capacity of the CCD, whether the camera could be mounted on an aeroplane and whether a wide angle lens could be used without distortion, Saner et al.[10]. A Panasonic EZI camera was found to fulfil the selection criteria.

The study then applied the GPS to the system to meet the requirements of the planned air video-graphs survey. A small hand GPS (Garmin-90) was found to be suitable for the study, which was then connected to a notebook computer running a navigation programme. The HYDRONAV programme was used for this aeroplane survey. The programme served two different purposes. It was used as graphical guidance for the pilot at different zoom levels and also recorded the routes with the time of the predefined frequency. The GPS information and images were combined via the time recordings of HYDRONAV.

The project's 1600 mosaic images were put together with the help of the constructed software. The goal was to present a general map composed of mosaic images of the area, so that individual images of the area could be produced with the click of a mouse.

All aspects of the system worked well. A small Cessna aeroplane carried out the survey. The study area was divided from north to south into 45 flight lines that were covered within 35 hours of flight. Each daily survey took approximately four hours of flight time and the total area was covered in 25 days. The plane flew at a height of 3,000 m during the study and the area covered by each individual image was 3.2 km x 4.8 km 10]. Even though only qualitative visual analysis was required, a quantitative analysis using automated computer based techniques was also prepared. A small area of the original study area, Tahtali Dam and reservoir, were chosen to determine the land use and crop pattern quantitatively.

The most significant aspect of this method is its economy, as it allows the acquisition of images at a very low cost. It could be a useful and inexpensive tool for other applications also.

3.4 COMPUTER-BASED COASTAL INFORMATION SYSTEM FOR THE TURKISH MEDITERRANEAN COASTS

Effective coastal management plans require the use of computer based information systems. In Turkey such a coastal information system is on the way. The aim of this new pilot project is to establish a system that will provide timely information to interested organisations, particularly government agencies. The integration of remote sensing data from different satellites with ground based data in a GIS is of particular importance in this project.

3.4.1 Study area
The Koycegiz-Dalyan Protection Area and its nearby surroundings situated on the Mediterranean coast have been selected as the study area for the pilot project. The area covers the Koycegiz, Ortaca, and Dalaman districts and has been declared a Special Protection Area by the Turkish Government, because of its importance as a nesting area for caretta caretta sea turtles, the occurrence of unique tree species and special ecological conditions, as well as for its historical past. As the area is both a coastal area and a protection area, the study will be challenging work for the participating scientists as it will require considerably sensitive and detailed data from many different sources, which will be integrated in a GIS. Figure 4 shows the Landsat-TM image of the study area.

3.4.2 Project goals
The goals of the project are as follows:

- To complete a pilot project involving the participation of scientists from different disciplines, different universities, and different countries using contemporary technologies.
- To make known the considerable pollution threats to the coastal areas, areas which are essential for Turkey but which have not, until now, been considered important enough to warrant the use of space technologies in their monitoring and management.
- To plan the extension of the coastal information system established in this project in the Koycegiz-Dalyan Protection Area and its surroundings, to the entire Mediterranean Coast.

78

- To promote co-operation between governmental organisations and universities, to facilitate the work of decision-makers and planners by supplying them with the findings, to use advanced technology and sophisticated science through specific applications, and to solve urgent problems and make recommendations for long-term problems. The involvement of governmental organisations is crucial to the success of the project.
- To develop joint projects with other Mediterranean countries in 2000 to extend the use of the coastal information system.

Figure 4. Landsat-TM image of the study area.

Organisations supporting the project: are the European Space Agency (ESA), Istanbul Technical University (ITU) and SPOT Image. Organisations participating in the project: are Istanbul Technical University (ITU), European Space Agency (ESA), Hacettepe University (HU), Russian Academy of Sciences (RAS) and Yildiz Technical University (YTU).

Disciplines of the project consist of remote sensing, coastal engineering, hydrogeology, geodesy, geology, surveying, urban planning, meteorology, hydrology and ecology.

3.4.3 Methods and data used:
Remote sensing.
Digital image processing.
- Processing of ERS ½ radar images.
- Processing of LANDSAT-TM images.
- Processing of SPOT-P images.
- Processing of KFA-1000 images.
- Processing of merged images.
- Image enhancement.
- Classification.

Digitising sheets and plans:
Ground Truth measurements
- GPS measurements.
- GPS supported water quality measurements.
- Preparation of bathymetric maps.
- Combination of GPS and echo sounder.
- Measurement of environmental parameters.
- Current measurements.
- Water level measurements.
- Dune monitoring.

Ancillary data
- Demographic.
- Meteorological.
- Ecological.
- Cadastral.
- Agricultural.
- Forestral.
- Tourist.

It is planned that the results of the project will be used by the Ministry of the Environment, the Ministry of Agriculture, the Ministry of Tourism, the Ministry of Culture, the Ministry of Public Works, the Authority of Special Protection Areas, the State Water Works, Provinces, Municipalities, etc.

4. Conclusions

Although there has been pressure to complete the project infrastructure promptly, particularly in the coastal areas of Turkey, financial constraints have slowed down the project. Great efforts have been made to construct the sewage collection systems, wastewater treatment plants and marine discharge systems of coastal provinces and districts. Priority items have been selected within the framework of a recent project, and construction work is still continuing. Marine discharge systems are preferred for those regions that have completed their sewage collection systems, leaving the construction of biological wastewater treatment plants to the long-term, again because of financial constraints. The Turkish coasts are among the less polluted coasts of the Mediterranean Sea; protective measures have to be taken as soon as possible to save the seas before they are considerably polluted. A recent international project on the development of a computer based coastal information system through integration of satellite and ground data is still on the way, and will lead to the development of effective coastal management plans.

5. References

1. UNEP (1993-1994) United Nations Environmental Programme. Environmental Data Report, Blackwell Publishers, Oxford, UK.
2. UNEP/ MAP (1996) United Nations Environmental Programme. The State of the Marine and Coastal Environment in the Mediterranean Region, MAP, Technical Reports, Series no: 100, Athens.
3. NEAP (1998) National Environmental Action Plan, State Planning Organisation, Ankara.
4. EFT (1995) Environmental Profile of Turkey, Environmental Foundation of Turkey, Ankara.
5. Samsunlu, A. (1996) Wastewater Treatment and Disposal at Coastal Regions of Turkey – Marmaris Example, First Uludag Environmental Engineering Symposium, Proceedings, p.1-19, 24-26 June 1996, Bursa –Turkey. (In Turkish).
6. Samsunlu, A., Akca, L., Uslu, O (1999). Wastewater Management at the Coastal Districts-Marmaris Example, Urban Management, Human and Environmental Problems Symposium '99, 17-19 February 1999, Istanbul, Proceedings, Vol. 3: Environmental Management and Control, pp.134-146. (In Turkish)
7. Samsunlu, A., Akca, L., (1999) Coastal Pollution and Mitigation Measures in Turkey, Water Science and Technology, Vol. 39, No: 8, 13-20.
8. Samsunlu, A., Akca, L., Uslu, O. (1995) Problems Related to an Existing Marine Outfall: Marmaris An Example, Water Science and Technology, Vol. 32, No: 2, 225-231.
9. Kuleli, T. (1999). Project on Coastal Zone Management and Tourism: Cirali and Belek, Development Report, July 1999, LIFE TCY96/TR/021, Association on the Preservation of Natural Life.
10. Saner, E., Eronat, A. H., Basoz, C., Uslu, O. (1999). An Economical New Approach to Airborne Videography- Case Study: Aegean Region of Turkey, Operational Remote Sensing for Sustainable Development, Nieuwenhuis, Vaughan & Molenaar (Eds), Balkema, Rotterdam, pp.113-117.

Part III: The Status of Water Resources Monitoring in the Balkans

ALBANIA

M. DELIANA
National Environmental Agency
Tirana, Albania

M. SANXHAKU, V. PUKA
Hydrometeorologic Institute
Tirana, Albania

L. SELFO
National Environmental Agency
Lake Ohrid Conservation Project
Ohrid, Albania

1. Introduction

Albania has entered a period of political and economic transition as it adopts a democratic form of government and a free market economy. It has inherited from the past an undeveloped economy and severe environmental problems. Damage to forests, loss of wetlands, overgrazing, agricultural development, mismanagement of terraces, outdated technologies, and the lack of wastewater treatment plants are among the many consequences of past neglect of natural resources and environmental degradation.

In the early 1990s economic and social change began a move towards a market economy. Increased foreign investments emphasised the need to address environmental protection issues. With few means of protecting the environment, the government gave priority to the strengthening of institutions and the establishment of a legal and regulatory framework. These in turn will create a stand from which economic and social activities may be addressed, in a spirit of harmony with the environment. Other priorities include urban waste management and biodiversity. Environmental issues are increasingly being taken into account by government ministries and research and scientific communities, and recently many water resource and environmental projects supported by the state or by international donors have been undertaken. Also the well known "polluter pay" principle has started to be applied in public and private sectors.

The role of various institutions has also been strengthened. The National Environmental Agency is directly under the Council of Ministers. Regional Environmental Agencies have been established at prefecture level, bringing improved management and assistance to the public and private sectors. These agencies contribute to the prevention of pollution and damage in new economic and social activities, and to solving existing environmental problems. A map of Albania's rivers is given in Figure 1.

J. Ganoulis et al. (eds.), Transboundary Water Resources in the Balkans, 81–85.
© 2000 *Kluwer Academic Publishers.*

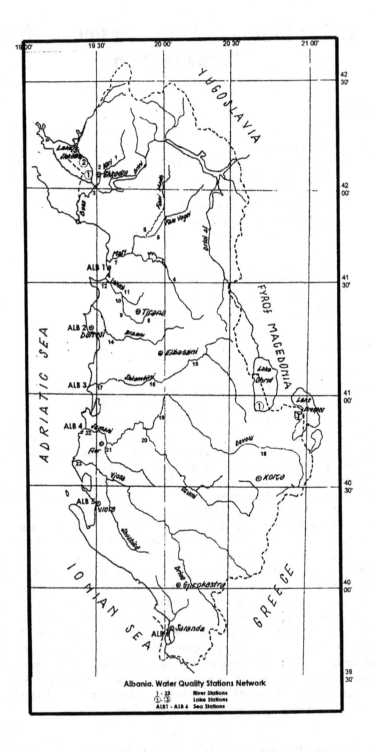

Figure 1. Albania's rivers, lakes and monitoring stations.

As well as the Council of Territory Adjustment at central and local levels, which deals with environmental issues, a number of other institutions are also involved, such as the Academy of Science, the National Committee of Water, the National Committee of Energy and the Committee of Tourism Development. The environmental activities initiated by the government are monitored by the Parliamentarian Commission on Health and Environment.

2. Institutional and Legal Framework for Environmental Protection

The Albanian parliament has approved a Law on Environmental Protection, whose purpose is the prevention and reduction of pollution, the conservation of biodiversity, the rational management of natural resources, and the ecological restoration of areas damaged by human activities or natural phenomena. The law contains special sections on environmental licenses, environmental impact assessment and monitoring.

A National Environmental Action Plan (NEAP) identifies priorities for environmental problems, in response to the new economic and social challenges Albania is now facing. In particular, NEAP sets environmental goals for relevant ministries. Albania is a signatory country to certain environmental conventions, such as the framework conventions on climate change, biodiversity protection, Barcelona, Bon and Ramsar conventions, etc. In 1992 the Albanian government adopted the Helsinki Convention on the Protection and Use of Transboundary Watercourses and International Lakes.

3. Water Monitoring

The Institute of Hydrometerology is mainly responsible for the monitoring of surface waters. A National Programme for Research and Development includes a Monitoring and Study on the Level of Pollution of Surface Waters in Albania and covers 17 rivers. The monitoring network comprises 23 river sampling stations, 5 in the lakes and 16 in coastal waters. Monitoring data are published in a bi-annual report on the environmental situation of the country prepared by the National Environment Agency (NEA).

3.1 RIVER MONITORING.

Parameters monitored in 1998 included temperature, pH, calcium, magnesium, sodium, potassium, hydrogen carbonate, carbonate, chlorides, sulphates, suspended solids, dissolved oxygen, chemical oxygen demand (COD), biological oxygen demand (BOD), nitrates, nitrites, ammonium, total phosphor, phosphates, etc.

From the values of the different parameters it is shown that Albanian river waters are slightly alkaline, having pH values of between 7.5 and 8.25 and have medium mineralisation (200-400 mg/l). There is an improvement in the content of suspended solids for some rivers (Uric, Fannie I Mad, Fannie I Vogel and Mat) compared to some years ago, when industrial plants were active and discharging wastewater with a high content of solid materials into these rivers.

Relatively high values of COD and BOD and other parameters in the rivers Ishmi and Lumi i Tiranes are explained by the high quantity of untreated sewage waters being discharged into these rivers from the city of Tirana (station 7 is located in the upper part of the river Ishmi). The quality of water in the Shkumbini River has also improved, because since 1990 there has no longer been any industrial discharge into these waters from metallurgical plants

Wastewater from oil exploitation activities is discharged into the river Gjanica, causing many parameters to deviate from their natural values. The same phenomenon is observed in the rivers Osum, Devoll and Seman and also caused by industrial activities, mainly oil processing. Nevertheless, most rivers maintain the natural characteristics. Their minerals content, COD, BOD, nitrogen and phosphor compounds are relatively low and dissolved oxygen has a level of saturation of 100%.

3.2 COASTAL WATERS

The monitoring of seawater is limited to the central part of coastal waters i.e. the Adriatic Sea. A monitoring project for the Adriatic Sea and the main rivers discharging into the sea was begun in co-operation with UNEP. The samples are taken in a coastal area between the mouth of the river Mati and Vlora bay. Higher levels of NBO at Durres, Vlora and Gryka e Ishmit stations are explained by the fact that untreated sewage water is discharged into the sea. The high content of suspended solids at some stations is linked to the content of solid materials in the river flow.

3.3 MONITORING OF LAKE WATERS

Seventy five per cent of Albanian lakes are shared with neighbouring countries. The monitoring of water quality in the lakes began only in the last few years. The frequency of sampling is once a year and the analysed parameters are the same as for river waters. Samples from lake Prespa are taken at the surface and at a depth 20 m. Samples for lake Ohrid are taken at the surface and at 4 different depths (25 m, 50 m, 75 m, and 100 m), and for lake Shkodra all samples are taken at the surface and near the bottom of lake (6-7 m). Generally speaking the quality of water in all three lakes is quite good, and the level of nutrients is low.

4. Future Programmes

A new Environmental Monitoring Programme is being prepared, which will help to achieve the objectives set by the NEAP. The new programme foresees measures for increasing the number of stations and parameters and the frequency of sampling. 25 sampling stations for monitoring river water with seasonal frequency are planned. The samples will analyse more than 20 parameters, including bacteriological parameters and at a few stations heavy metal and hydrocarbons as well.

As far as lakes are concerned monitoring parameters will be extended to include biota and sediments. Samples will be taken every season, but only annually for biota and sediments. Monitoring of the sea will be extended beyond the actual limitations, covering both Adriatic and Ionian Albanian coastal waters, and the number of parameters will be increased to include biota and sediments.

For the first time water monitoring will also cover reservoir waters used for irrigation, and ground waters in some areas with more intensive industrial and agriculture activities, mainly in the western lowland. Data collection and management will be upgraded by introducing contemporary methods.

5. Additional Data and Information

5.1 POLICY INFORMATION

- The Ministry responsible for monitoring water quantity and quality is the National Environmental Agency (NEA), contact Maksim Deliana, Blvd zhan d'Arc, Tirana.
- Primary legislation: the Law on Environmental Protection, 1993 plus regulations, law on state Sanitary Protectorate, 1992.
- Agencies developing and enforcing standards: NEA and Ministry of Health, contact Maksim Deliana, zhan d'Arc, Tirana.
- Scientific review by scientists and other specialists: Provided by specialists from the Institutes of Academy of Science, universities and other institutes.
- Recommendations: An adequate monitoring network for the whole country.

5.2 DATA AND INFORMATION ON TRANSBOUNDARY MONITORING

Albania is bordered by Yugoslavia, the Former Yugoslav Republic of Macedonia (FYROM) and Greece. It shares rivers as follows: the Drini with FYROM, the Vjosa with Greece and the Bura with Yugoslavia. Lakes are shared as follows: Ohrid with FYROM, Prespa with FYROM and Greece, and Shkodra with Montenegro
With respect to river monitoring, at the Drini and Vjosa sites there are 3 stations. The following parameters are monitored bi-monthly: pH, temperature, Ca, Mg, Na, K, HCO_3, SO_4, NO_2, NO_3, SiO_2, P_2O_5, Fe, and additionally since 1984:O_2, COD, NH_4, Zn, Cu, Mn. Lake Ohrid is monitored as follows: temperature, NO_3, NO_2, NH_4, PO_4, P total, SiO_2, BOD, COD twice a year.
Albania has no monitoring agreements at present although it has signed certain relevant international conventions. Recommendations include: The establishment of a new network for the Institute of Hydrometeorology is proposed, with four stations in Drini and Vjosa, with bi-monthly sampling frequency. Also a new network of three stations is suggested for Lake Ohrid, to be monitored at different depths, with the additional parameter of transparency. Two stations are proposed for Lake Prespa and three for Lake Shkodra, both with a frequency of bi-monthly sampling.

6. References

1. Water quality monitoring in Albania, Technical Report, 1996, HIDMET, Tirana
2. Water quality monitoring in Albania, Technical Report, 1997, HIDMET, Tirana
3. Water quality monitoring in Albania, Technical Report, 1998, HIDMET, Tirana.

BOSNIA AND HERZEGOVINA

T. KUPUSOVIC
Hydro-Engineering Institute
Sarajevo, Bosnia and Herzegovina

1. Introduction

The highest regions of the central part of the Dinaric mountains of Bosnia and Herzegovina (B&H) are located in the Danube basin watershed. The inner part of this watershed slopes gently inland, whereas the outer falls away steeply and drains directly into the Adriatic Sea. As much as three quarters of B&H (area of 51,129 km^2 in total) is drained by rivers that flow into the river Sava, a tributary of the river Danube; the remaining quarter belongs to the Adriatic sea basin.(Figure 1).

The river Sava flows through the north of B&H and forms the boundary between B&H and Croatia. The rivers Una and Drina are also transboundary rivers, forming boundaries with Croatia and Yugoslavia respectively. In the Mediterranean part of B&H, only the river Neretva, which rises beyond karst territory, manages to pass through the karst mountains.

The war of 1992-95 and the political changes that followed reflect the present state of water quality and quantity monitoring in B&H. Presently there is no unified monitoring service on a state level, and practically no systematic monitoring at all. Responsibilities among ministries are not clearly defined. Existing laws, the present Water Law (May 1998) and legislation from the pre-war period still in force are not applied and do not comply with European Union (EU) regulations.

The PHARE project "Water Sector Institution Strengthening" (Federation of B&H) is already underway. The same project for the Republic of Srpska is expected to start soon. Its main thrust will be to establish completely new water and environment agencies, replacing obsolete organisations. The enactment of water and environment legislation compliant with EU regulations is anticipated.

As a Danube River basin country with the intention of becoming a signatory country to the Danube River Protection Convention, B&H will play an active role in the Environmental Programme for the Danube River Basin. Four stations are planned for stage 1 of the TransNational Monitoring Network (TNMN) to be located in B&H. During 1998 B&H also participated in activities under the Convention on Protection and the Use of Transboundary Watercourses and International Lakes, again as a country with the intention of becoming a signatory.

J. Ganoulis et al. (eds.), Transboundary Water Resources in the Balkans, 87–97.
© 2000 *Kluwer Academic Publishers.*

88

Figure 1 Rivers of Bosnia and Herzegovina.

Since the river Neretva is a transboundary river flowing through B&H towards Croatia, it is covered by the Mediterranean Hydrological Cycle Observing System (MED-HYCOS). This system aims to improve co-operation between the countries of the Mediterranean Sea basin in the field of water resources assessments and management, including the collection, transmission, processing and dissemination of relevant water data

The year 1998 saw intensive activities under the Mediterranean Action Plan (MAP) Office of B&H resulting from the development of very close co-operation between scientific institutions from B&H and Croatia, as well as with scientists and institutions from other countries.

2. State of Water Monitoring in B&H

2.1 ORGANISATION OF MONITORING BEFORE THE WAR

2.1.1. *Legal Regulations*
Under the 1965 framework of water legislation it was stated that hydrometeorological services were responsible for the systematic monitoring of water quantity and quality in the Republic of B&H. Later these hydrometeorological services operating on behalf of the Republic were uniformly regulated, using the same methodology and standards defined for the whole of the former Yugoslavia. The Hydrometeorologic Institute of the former Yugoslav Federation followed the criteria and guidelines of the World Meteorological Organisation (WMO), since the Federation was a member.

Additional legislation was adopted in the late 1980s that improved the monitoring network design and the programme of the hydrological stations This included systematic monthly observation of water quantity and quality elements. The next logical step would have been the harmonisation of laws in the Socialistic Republic of B&H with those in former Yugoslavia, however this did not occur because of political tensions and the impact of the war, which followed. All of the above mentioned law and bylaw regulations are still in force in B&H, under regulations adopted in 1992.

2.1.2. *Monitoring Network*
The water quantity and quality monitoring network in operation before the war included 371 stations for the monitoring of surface water quantity parameters.144 stations were included in the basic network. At 6 stations, suspended solids load measurements were also being monitored, and at 44 stations water temperature was also being monitored. On average, the stations had been functioning for 30 years and those with level recorders for 15 years. Discharge measurements were also being done, averaging 1.6 per station per annum.

Systematic observation of B&H surface waters quality began in 1965. The national statutory (basic) monitoring network of the Hydrometeorologic Institute of B&H covered B&H rivers with a total of 58 stations, 53 of which were situated in the Black Sea (Danube River) basin, and 5 in the Adriatic Sea basin. The area covered by sampling sites varies from an average of 730 km^2 per sampling site (Black Sea basin) to an average of 2482 km^2 per sampling site in the Adriatic Sea basin. This difference is caused by differences in river network density. In the Mediterranean part of B&H, only the 233 km long Neretva River, which rises beyond karst territory, manages to pass through the karst mountains.

2.1.3. *Methods of Water Quality Data Collection*

The monitoring and evaluation of water quality was based on occasional grab water samples, Chemical parameters were monitored three times a year (spring, summer and autumn), and biological parameter twice a year (summer and autumn). Standard physical-chemical parameters, such as temperature, appearance, pH-value, alkalinity, dissolved oxygen and degree of saturation, and afterwards all types of hardness, total and suspended solids, $KMnO^4$ consumption, BOD^5, ortho-phosphate and total iron were determined in all grab samples, Nitrogen compounds (ammonia, nitrite and nitrate nitrogen) were checked at ten stations only. Heavy metal analyses were performed from time to time using instruments that ensured a general type of analysis, but did not provide reliable date on characteristic concentration for watercourses.

Other specifications, particularly for organic pollutants, including a wide range of organic micro-pollutants standards (pesticides and other biocides) were not included in regular control. Indeed the Republic Hydrometeorologic Institute does not have the necessary equipment for such determinations. However, biological and sanitary bacteriological parameters of water quality were regularly determined at the stations. Groundwater, except in the Semberija region and stagnant water (water and reservoirs) was not included in the regular programme of water quality control.

Monitoring conditions before the war were not satisfactory for quantitative or qualitative components. Documents issued just before the war, such as: "The Programme of Modernisation and Technical Equipping of Hydrometeorologic Service in B&H for the period 1991-2000" (1990), "Conception of Long-term Programme of Water Protection"(1991), "Regulation on Network Establishment and Operation Programme for Observation Stations" (1990) were unsuccessful attempts to improve water monitoring conditions in B&H.

2.2 TRANSBOUNDARY ASPECT

Some of the surface water quantity and quality monitoring stations in B&H, which were in operation before the war, were of transboundary importance. According to the 1986 decision on the development of stations (Odluka o osnovnoj mrezi hidrometeoroloskih stanica - 51.1ist SRB&H br.6/86), inter alia, the two criteria to be met were:

- An international warning station.
- An inter-republic warning station. (The term inter-republic refers to the relationship between the six republics of the former Yugoslav federation.)

All of the seven stations under criterion 1 were situated in the Danube River basin. The former Yugoslavia was a signatory country to the convention on safety navigation on the Danube River and was obliged to submit daily water level data from these stations to the Federal Hydrometeorologic Institute in Belgrade. Of the 27 stations under criterion 2, three were located in the Adriatic Sea basin.

Of the 11 stations on the transboundary river Sava, only two (Brcko and Bosanska Raca) came under the responsibility of the B&H hydrometeorologic service. Some stations on B&H rivers flowing into the river Sava (station Kostajnica - River Una, station Derventa - River Ukrina, station Delibasino Selo - River Vrbas) were automatic and included in the system for telemetric reading of water levels and remote control of the river Sava flood regulation structures. This system was constructed and funded by Croatia. The VAX computer in Croatian Waters and the hydrometeorological service enabled modem communications system to be established. It was intended that a similar system of communication should be set up with services in Sarajevo. The gauging stations of this system can serve eight channels, or eight measuring devices, however, so far only one has been used, that of measuring water levels.

In the regulations on water quality monitoring it is stated that: "where water courses are cut by a state boundary, a station should be set up in a boundary zone, of up to 10 km, and where water courses form a state boundary, a station should be set up at a distance of between 50 and 70 km." It was also stated that stations should be set up "at the part of the water course upstream of the mouth, whether flowing into a main course, lake or sea". Of the 58 network stations for water quality control in B&H, 14 were located at the mouth, but no sampling was done in any of the tributaries of the Sava or Neretva rivers.

3. Present State of Monitoring

3.1 WATER QUANTITY

With the outbreak of war, monitoring under B&H's systems of water quantity and quality stopped altogether. B&H's new political organisation is reflected in its hydrometearologic service. The Federal Meteorological Institute, located in Sarajevo, is now responsible for the territory of the Federation of B&H, and the Hydrometeorologic Institute situated in Banja Luka, is active in the territory of the Republic of Srpska. The Meteorological Institute of B&H is a member of the World Meteorological Organisation but still has no employees.

Surface water quantity elements are observed at 27 profiles/stations, including 15 profiles equipped with water level recorders. The activities during the war of 1992-95 and the decentralisation of hydrometeorologic services, resulting from the changed political organisation of B&H, have produced significant differences in water quantity and quality observation.

The Public Company for the B&H watershed of the Adriatic Sea and Croatian water management authorities have organised and funded the development of a hydrological network in the B&H Adriatic Sea watershed, with equipment supply, maintenance and data processing. Of the total number of stations active in the Federation, 22 are located in the Mediterranean part (Adriatic Sea basin), and there are 14 water level recorders.

3.2 WATER QUALITY

Water quality monitoring is performed at only 13 sites in B&H. The Hydrometeorologic Institute, Republic of Srpska, observes the quality of water in the river Vrbas at 8 profiles. The Public Company for the B&H watershed of the Adriatic Sea has organised monthly sampling at sites on the Neretva River.

In the Danube River basin part of the Federation, there is no monitoring of surface water quality. In December 1998, under the Federal Meteorological Institute, one automatic station with remote data transfer was installed on the river Bosnia near Sarajevo, as a pilot station of water management information system. This station observes water level and has 5 water quality parameters (temperature, conductivity, salinity, pH-value, redox-potential and dissolved oxygen).

Under existing Water Law, public companies in the watersheds are responsible for the organisation and practise of water flow regime observation in the territory of the Federation. Since the regulations of the pre-war period are still in force, there are many overlapping areas of responsibility and obscurities in legal interpretation, because before the war the hydrometeorologic service was responsible for monitoring. This was a government decision and was financed accordingly. In the meantime, there is really no systematic monitoring.

3.3 NEW WATER MANAGEMENT PRINCIPLES

The PHARE project. "Water Sector Institution Strengthening" (Federation of B&H) is already underway. Its proposed water managementprinciples have been widely adopted by the EU and other international agencies.

The key principles in water management, suggested for the Federation of B&H are as follows:

- Water protection to be integrated into environmental administration.
- Licensing to be separate from all other functions.
- Research and development to be separate from regulation.
- Subsidies to be minimised to the lowest possible level.
- Administration to be financed from relevant budgets, not directly from levies, user charges or other revenues
- Services and economic activities to be owned and managed by users/beneficiaries/service providers.
- Promotion of specific sub-sectors by respective line ministries.

The main principle in the implementation is the establishment of completely new water and environment organisations to replace those that are now obsolete. The profound changes required by the new institutional set-up and prerequisites of the EU and other international rules and principles, combined with the defects of the present Water law, make entirely new legislation in the water sector necessary. Enactment of water and environmental legislation compliant with EU regulations is one of the short-term actions that is anticipated.

The same project is expected to be set up in the Republic of Srpska. At the same time, the harmonisation of institutional strengthening with the Republic of Srpska is to come under the responsibility of the B&H Environment Steering Committee.

4. Transboundary Issues

B&H is included in the Danube River Basin Environment Programme, as a Danube basin country with the intention of becoming a signatory country to the Danube Convention, as well as a signatory to the Convention on the Protection and Use of Transboundary Watercourses and International Lakes.

Ample EU PHARE Multi-Country Environmental Programme funding has been made available to B&H, so that it may comply with the requirements of the Danube River Protection Convention with regard to the Monitoring, Laboratory and Information Management (ML.IM) programme and the Accident Emergency System (AEWS). MLIM and AEWS will be operated and maintained in an integrated manner for both the Federation of B&H and the Republic of Srpska, with four modem stations, a National Reference Laboratory (NRL) and a Principal International Centre (PIAC). Experts from the Danube countries have been involved in the development of the TransNational Monitoring Network (TNMN) focused on establishing a monitoring network for water quality data through the river basin. The monitoring network shall provide outputs compatible with other major international river basins in Europe, and in time will comply with standards used in the western part of Europe-

All stations of the network were given the same minimum sampling frequency of 12 per year for determinants in water, and 2 per year for determinants in sediment and bio-monitoring. The number of riparian countries participating in the working groups during implementation were increased from 8 to 12 (Moldova, Germany, Austria and B&H) which led to an increase in the number of stations to 61. Stations to be located in B&H planned for phase 1 of the TNMN are presented in Table 1

TABLE 1. Stations in B&H for phase 1 of the TNMN.

Country Code	River Name	Town/Location	Longitude d.m.s.	Latitude d.m.s.	Distance	Altitude	Catchment	Loc. Profile
B&H01	Sava	Jasenovac	45 16 0	16 54 36	500	87	38953	M
B&H02	Sava/ Una	Kozarska Dubica	45 11 6	16 48 42	16	94	9130	M
B&H03	Sava/Vrbas	Razboj	45 03 36	17 27 30	12	100	6023	M
B&H04	Sava/ Bosna	Modrica	44 58 17	18 17 40	24	99	10308	M

The lists of determinants for water and sediments as agreed for TNMN Phase I is presented in Table 2 and Table 3 in the Annex.

The Hydrometeorologic Institute of the Republic of Srpska supplies daily data on water quantity elements related to 8 profiles/stations located on the Sava River and its tributaries (station Srbac/River Sava, Gradiska/River Sava, Raca/River Sava, Novi Grad/River Una, Prijedor/River Sana, Banja Luka/River Vrbas, Doboy/River Bosna and Foca (Srbinje)/RiverDrina).

The Neretva, the main river in the Mediterranean part of B&H, is covered by the Mediterranean Hydrological Cycle Observing System (MED-HYCOS), a regional component of World-HYCOS. Of the activities proposed within the framework of the Implementation Programme of the MED-HYCOS Project, the first immediate objective is the installation of a network of key stations and multisensor-equipped Data Collection Platforms (DCPs) for the collection and transfer of several variables related to water resources monitoring. There is not yet a single DCP on the B&H part of the Neretva River. Croatia, also a Mediterranean country, chose to locate a MED-HYCOS DCP on the Neretva River at the profile at Metkovic.

5. Scientific Support and Conclusions

It is clear that the management of complex water resources cannot take place in the absence of scientific support. The deterioration in water quality in some B&H regions was a critical problem faced by the country's water management authorities before the war. This situation developed primarily as a result of inappropriate management and ignorance of the consequences, and partly because of inadequate information and knowledge. The early nineties saw significant changes in the global principles associated with water resources management. The principles that have been identified and highlighted at a number of International meetings include:

- Identification of water as an economic good with an associated economic value in all of its competing uses.
- Importance of water within the principle of ecologically sustainable development.
- Multidisciplinary nature of water resources management and recognition of freshwater as a scarce and vulnerable resource, essential to the preservation of all forms of life.

At this stage of water management in B&H, which is focusing on the adoption of new water management principles and the establishment of new institutions, scientists and managers should work together as a team to insure that scientific information is relevant and properly applied for management purposes. Optimally, scientists are vital to the understanding of the functioning of nature, and to the recognition of the importance of actual global guiding principles. Scientific support is needed in the selection of management control measures and in the preparation of materials for public information and education.

In the "Water Sector Institution Strengthening", project started in June 1998, professors and assistants from the Hydro-Engineering Institute Faculty of Civil Engineering of Sarajevo participated as local team consultants. They are currently working on the water quality aspect of the water sector institutional set up. Water quantity and water quality studies will be essential in the process of the development of the B&H monitoring network.

Doctors of sciences, engineers, professors and assistants are engaged as representatives of B&H in the Environmental Programme of the Danube River Basin for activities under the Convention on the Protection and Use of Transboundary Watercourses and International Lakes, and for activities of the Mediterranean Action Plan (MAP) Office. With the aim of the development and improvement of environmental practice in B&H, representatives of institutions involved in MAP activities have attended various meetings and training courses organised under different programmes.

1998 saw intensive activities under the MAP Office of B&H, resulting from the development of very close co-operation between scientific institutions from B&H and Croatia, as well as with scientists and institutions from other countries. On the list of local and foreign experts and institutions involved in project implementation are professors from different faculties in B&H and Croatia.

Annex

TABLE 2. Determinants list for water for phase 1 of the TNMN.

Determinants in water	Unit	Minimum level of interest	Principal level of interest	Target limit of detection	Tolerance
Flow	m^3/s				
Temperature	^0C	-	0-25	-	0.1
Suspended Solids	mg/l	1	10	1	1 or 20%
Dissolved Oxygen	mg/l	0.5	5	0.2	0.2 or 10%
pH	-	-	7.5	-	0.1
Conductivity 20^0 C	?S/cm	30	300	5	5 or 10%
Alkalinity	mmol/l	1	10	0.1	0.1
Ammonium(NH_4^+)	N mg/l	0.05	0.5	0.02	0.02 or 20%
Nitrite (NO_2^-)	N mg/l	0.005	0.02	0.005	0.005 or 20%
Nitrate (NO_3^-)	N mg/l	0.2	1	0.1	0.1 or 20%
Organic Nitrogen ($KjN - NH_4^+$)	N mg/l	0.2	2	0.1	0.1 or 20%
Total Nitrogen ($KjN + No_x$)	N mg/l	0.2	2	0.5	0.5
Soluble Reactive Phosphate	P mg/l	0.02	0.2	0.005	0.005 or 20%
Total Phosphorus	P mg/l	0.05	0.5	0.01	0.01 or 20%
Sodium	mg/l	1	10	0.1	0.1 or 10%
Potassium	mg/l	0.5	5	0.1	0.1 or 10%
Calcium	mg/l	2	20	0.2	0.1 or 10%
Magnesium	mg/l	0.5	5	0.1	0.2 or 10%
Chloride	mg/l	5	50	1	1 or 10%
Sulphate	SO_4^{2-}mg/l	5	50	5	5 or 20%

Determinants in water	Unit	Minimum level of interest	Principal level of interest	Target limit of detection	Tolerance
Iron	mg/l	0.05	0.5	0.02	0.02 or 20%
Manganese	mg/l	0.05	0.5	0.01	0.01 or 20%
Zinc	?g/l	10	100	3	3 or 20%
Copper	?g/l	10	100	3	3 or 20%
Chromium	?g/l	10	100	3	3 or 20%
Lead	?g/l	10	100	3	3 or 20%
Cadmium	?g/l	1	10	0.5	0.5 or 20%
Mercury	?g/l	1	10	0.3	0.3 or 20%
Nickel	?g/l	10	100	3	3 or 20%
Arsenic	?g/l	10	100	3	3 or 20%
Aluminium	?g/l	10	100	10	10 or 20%
BOD_5	mg/l	0.5	5	0.5	0.5 or 20%
COD_{Cr}	mg/l	10	50	10	10 or 20%
COD_{Mn}	mg/l	1	10	0.3	0.3 or 20%
DOC	mg/l	0.3	1	0.3	0.3 or 20%
Phenol index	mg/l	0.005	0.05	0.005	0.005 or 20%
Anionic surfactants	mg/l	0.1	1	0.03	0.03 or 20%
Petroleum hydrocarbons	mg/l	0.02	0.2	0.05	0.05 or 20%
AOX	mg/l	0.01	0.1	0.01	0.01 or 20%
Lindane	?g/l	0.05	0.5	0.01	0.01 or 30%
pp'DDT	?g/l	0.05	0.5	0.01	0.01 or 30%
Atrazine	?g/l	0.1	1	0.02	0.02 or 30%
Chloroform	?g/l	0.1	1	0.02	0.02 or 30%
Carbon tetrachloride	?g/l	0.1	1	0.02	0.02 or 30%
Trichlorethylene	?g/l	0.1	1	0.02	0.02 or 30%
Tetrachlorethylene	?g/l	0.1	1	0.02	0.02 or 30%
Total Coliforms (37 C)	CFU/100ml	-	-	-	-
Faecal Coliforms (44 C)	CFU/100ml	-	-	-	-
Faecal Streptococci	CFU/100ml	-	-	-	-
Salmonella sp	in 1 litre	-	-	-	-
Macrozoobenthos	no.of taxa	-	-	-	-
Macrozoobenthos	sapr.index	-	-	-	-
Chlorophyll A	?g/l	-	-	-	-

TABLE 3. Determinants list for sediments for phase 1 of the TNMN.

Determinants in sediments	Unit	Minimum likely level of interest	Principal level of interest	Target limit of detection	Tolerance
Organic Nitrogen (KjN - NH$_4^+$)	N mg/kg	50	500	10	10 or 20%
Total Phosphorus	P mg/kg	50	500	10	10 or 20%
Calcium	mg/kg	1000	10000	300	300 or20%
Magnesium	mg/kg	1000	10000	300	300 or 20%
Iron	mg/kg	50	500	20	20 or 20%
Manganese	mg/kg	50	500	20	20 or 20%
Zinc	mg/kg	250	500	50	50 or 20%
Cooper	mg/kg	2	20	1	1 or 20%
Chromium	mg/kg	2	20	1	1 or 20%
Lead	mg/kg	2	20	1	1 or 20%
Cadmium	mg/kg	0.05	0.5	0.05	0.05 or 20%
Mercury	mg/kg	0.05	0.5	0.01	0.01 or 20%
Nickel	mg/kg	2	20	1	1 or 20%
Arsenic	mg/kg	2	20	1	1 or 20%
Aluminium	mg/kg	50	500	50	50 or 20%
TOC	mg/kg	50000	500000	10000	10000 or 20%
Petroleum hydrocarbons	mg/kg	10	100	1	1 or 20%
Total Extractable matter	mg/kg	100	1000	10	10 or 20%
PAH - 6 (each)	mg/kg	0.01	0.1	0.003	0.003 or 30%
Lindane	mg/kg	0.01	0.1	0.003	0.003 or 30%
pp'DDT	mg/kg	0.01	0.1	0.003	0.003 or 30%
PCB - 7 (each)	mg/kg	0.01	0.1	0.003	0.003 or 30%

6. References

1. Water Management Institute: Conception of Long-Term Water Protection, Sarajevo 1991.
2. Hydrometeorological Institure of SR B&H: Regulation on Network Establishment and Operational Programme for Observation Stations, Sarajevo 1990.
3. 1996-1998 The MLIM Subgroup:Water Quality in the Danube River Basin, TNMN Yearbook 1996.

BULGARIA

R. ARSOV
G. MIHAILOV
University of Architecture
Sofia, Bulgaria

G. MIRINCHEV
Ministry of Environment and Water
Sofia, Bulgaria

1. Introduction

Almost all rivers in Bulgaria rise in Bulgarian territory, see Figure 1. The only exceptions are the very upper reaches of four small rivers, the Erma and Dragovishtitza (rising in Yugoslavia), the Strumeshnitza (rising in the Former Yugoslav Republic of Macedonia (FYROM)) and the Rezovska (rising in Turkey). The reaches of three rivers form part of the Bulgarian state borders, namely the Danube (with Romania), the Timok (with Yugoslavia) and the Rezovska (with Turkey) and part of the Maritza (Evros, Merich) River, which rises in Bulgaria, forms the state border between Greece and Turkey. The 35 separate river sub-basins in Bulgaria belong in general to two main catchments, separated by the Balkan mountain chain. The first of them is the Black Sea catchment, which includes the northern Bulgarian rivers, entering indirectly through the Danube River, as well as all the rivers entering directly to the sea. The second main catchment is the Aegean Sea, to which the most of the southern Bulgarian rivers belong.

The rivers Struma, Dospat, Mesta, Maritza, Arda and Byala cross the state border between Bulgaria and, and, another, the Tundja, crosses the state border between Bulgaria and Turkey. The Arda River also crosses the Greek-Turkish border, flowing into the Maritza River near the town of Edrine. For years these seven rivers have been of special interest, importance and concern for the three countries concerned in term of both water quantity and quality. International activities and co-ordination in this field have a long history. Some general data are given in Table 1 for the southern Bulgarian river basins under consideration.

Since the northern Bulgarian rivers affect mainly the multinational catchments of the Danube River and the Black Sea, their monitoring and impact is better discussed at already well established forums, such as the Danube Programme and the Black Sea Programme, with all the states concerned, including non-Balkan countries.

Since a water quality monitoring network forms part of a water resources management system, a brief review of the institutional aspects of Bulgaria's current water resources management system follows.

J. Ganoulis et al. (eds.), Transboundary Water Resources in the Balkans, 99–105.

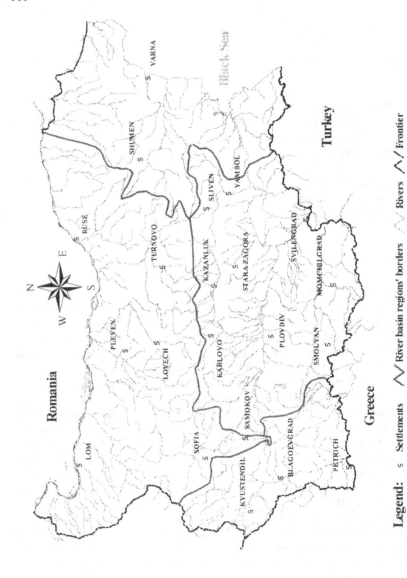

Figure 1. River network and river basin regions in Bulgaria.

Legend: s Settlements ∧ River basin regions' borders ∧ Rivers ∧ Frontier

TABLE 1. An Overview of Southern Bulgarian River Basin Characteristics.

Name of river in Bulgarian	Name of river in Greek	Catch- ment area km^2	Population (1995)	Ann- ual flow m^3	River length km	
					Bulgaria	Greece
1 Struma	Strymon	10,250	542,100	2,466	290	110
2.1 Mesta	Nestos	2,770	137,000	1,120	126	120
2.2 Dospat	Nestos	630	included	235	96	20
3.1 Arda	Arda	5,200	323,600	2,204	240	30
3.2 Bjala	Erithropo tamos	590	included	192	60	30
4 Maritsa	Evros	21,000	1,758,000	4,319	321.5	160
5 Tundja		7,780	520,900	1,080	350	0

2. Water Quality Management – A Brief Review

2.1 INSTITUTIONAL ASPECTS

Water management in Bulgaria is the responsibility of several different ministries and other organisations led mainly by the Ministry for Environment and Water, the Ministry of Regional Development and Public Works, the municipalities, and the water companies. The exact responsibilities and authorities of these institutions are defined in Bulgaria's environmental legislation.

2.1.1 Ministry of Environment and Water
The Ministry of Environment and Water (MEW) is responsible for the environmental protection and use of natural resources including water. It is the responsibility of MEW to:

- Establish national environmental policy and programmes.
- Propose legislation.
- Approve standards, norms, charges and the level of fines.
- Accept projects, expertise and impact assessments.
- Organise and manage the National Monitoring system.
- Co-ordinate and execute the state control on environmental protection.

2.12 Ministry for Regional Development and Public Works
There are two relevant departments in the Ministry for Regional Development and Public Works (MRDPW), comprising the Water Sector and the PHARE Project Management Unit (PMU), plus the local service units based on municipalities or water companies.

2.2 LEGISLATIVE ASPECTS

The main legislation still in force during the transitional period until the end of 1999 in the Water Sector in Bulgaria is as follows:

- Law on Air, Water and Soil Protection from Pollution (1963, supplemented and changed in 1968, 1969, 1975, 1977, 1978, 1988, 1991 and 1992); and several regulations.

- Environment Protection Law (1991). This law concerns information collection, control and assessment of the environment. Regulation No 1 on Environmental Impact Assessment (1995.

- Water Law (1969, supplemented and changed in 1977, 1979, 1984, 1986 and 1987). This law sets up the rules for issuing water permits, defines water ownership, provides a statement for the organisation and function of the National Water Council (later merged with the Ministry of Environment and Water)

The new Water law has been approved by the Council of Ministers and was voted in by Parliament on 2 July 1999 with a transitional period of 6 months. The new Water law envisages a new integrated scheme of water management in Bulgaria, see Figure 4. It will impose new responsibilities and restructure the Ministry of Environment and Water. This is reflected in a new river basins' structure arrangement where basins are grouped in four regions, as shown in Figure 1.

6. Water Quality Monitoring – A Brief Review

3.1 BACKGROUND

The river-monitoring network in Bulgaria was established in 1893 and has since been expanded and rearranged several times. Initially only water quantity was measured. The National Institute of Meteorology and Hydrology (NIMH) at the Bulgarian Academy of Science (BAS) is currently responsible for national surface water quantity monitoring system management.

Since 1976 and until recently, two water quality monitoring networks operated in parallel on Bulgarian rivers and some reservoirs. One network is operated by the NIMH and one by the MEW. The monitoring system at MEW however, serves as a base for water quality management, while the one at the BAS plays a supplementary role. It is curious to note that the data from the monitoring system at the BAS (which is funded by the state) are now only available on a commercial basis, even to the state services.

In April 1998, by order of the Minister of the MEW, a new working group was formed at the MEW and new monitoring operations were developed. The water quality parameters, which differ to those at ecological gauging stations, are given in Table 2. The recommended sampling frequency is as follows:

- Gauging stations of kind B - 4 times a year, in February, May, August and November.
- Gauging stations of kind E - twice a year in March and September.
- Gauging stations of kind I, C, M and R - once a month.

The water quality analyses are according to the Bulgarian State Standard (BSS), ISO and EN. The proposed new arrangement of the NSEM came into force in July 1998.

A number of ecological gauging stations have been determined in the southern Bulgarian transboundary rivers. In addition, three fully automated gauging stations with continuous sampling have been delivered recently through a project funded by the PHARE Programme. They are located in the river basins of the Struma (Strimon), Mesta (Nestos) and Maritza (Evros, Merich) and collect data on accidental pollution and warn of possible transboundary pollution.

TABLE 2. Water quality parameters, defined at the ecological gauging stations.

Type of sampling points	Parameters	
	Main	Supplementary
Background – B	Temperature, pH, O_2, conductivity, MnO_4-COD, BOD_5, N_{tot}, N_{org}, NO_3, NO_2, PO_4, SS	Mn, Fe
Intermediate – I	Temperature, pH, O_2, conductivity, MnO_4-COD, BOD_5, N_{tot}, N_{org}, NO_3, NO_2, PO_4, SS, COD, Cl, SO_4	Heavy metals, detergents, phenols, CN, oil derivatives, Cl_{org}, pesticides, polyaromatic carbohydrates
Mouth - M	Temperature, pH, O_2, conductivity, MnO_4-COD, BOD_5, N_{tot}, N_{org}, NO_3, NO_2, PO_4, SS, COD, Cl, SO_4	Heavy metals, detergents, phenols, CN, oil derivatives, Cl_{org}, pesticides, polyaromatic carbohydrates
Cross-boundary – C	Temperature, pH, O_2, conductivity, MnO_4-COD, BOD_5, N_{tot}, N_{org}, NO_3, NO_2, PO_4, SS, COD, Cl, SO_4	Heavy metals, detergents, phenols, CN, oil derivatives, Cl_{org}, pesticides, polyaromatic carbohydrates
Reservoirs & lakes - R	Temperature, pH, O_2, conductivity, MnO_4-COD, BOD_5, N_{tot}, N_{org}, NO_3, NO_2, PO_4, SS	Heavy metals, detergents, phenols, CN, oil derivatives, Cl_{org}, Pesticides, polyaromatic carbohydrates
Emissions - E	According to specific schemes (individual)	

4. Scientific Support for Water Quality Monitoring System Development

Scientists and experts from the national research institutions formally provide scientific support for the NSEM development. In reality scientific support for water quality monitoring and management is much wider, since many scientific and research projects have been conducted on this subject for years. Some of them are referenced below.

4.1 RECENT SCIENTIFIC AND RESEARCH PROJECTS

4.1.1. Problems
The problems associated with the development and improvement of the NSEM are not of a purely scientific or purely operational nature. However, an overall scientific approach is obviously indispensable, and with this is mind the following problems can be mentioned:

- Lack of enough contemporary laboratory equipment for all the necessary water analyses at the RIEW's laboratories:

- The existence of a complicated relationship with co-ordination difficulties between the three institutions responsible for determining river water quantity, namely the RIEWs at the National Centre for Environment and Sustainable Development (NCESD) (MEW), the NIMH at BAS and the Department of Reservoirs and Cascades (DRC) at the Ministry of Agriculture and Forestry (MAF). The latter is responsible for the management of large reservoirs and river flow rates around the reservoirs.

- Lack of an adequate information system for water quantity/quality data storage, processing and retrieval, capable of producing reports for decisions making and operational activities.

- Lack of a reliable system for incidental pollution registration and early warning, based on a network of gauging stations with continuous (24 h) sampling and quickly produced analyses.

Special attention should be given to problems associated with specific river water quality monitoring. This is necessary for calibration and verification of the relevant simulation models. The latter are indispensable nowadays for the adequate determination of the assimilative capacity of water bodies for pollutants of different kinds. The mathematical modelling of river water quality dynamics forms the basis modern river basins management technology. Obviously a specific approach is needed for the development and maintenance of such a monitoring system, based on strong scientific criteria and relevant systematic procedures, distributed in space and time. This issue is still not well clarified in world current practice, despite the availability of relevant technology.

5. Current Status of Water Quality of the Southern Bulgarian Trans-boundary Rivers

The previous national institutions responsible for water management (later replaced by the Ministry of the Environment and Waters) adopted Water Quality Ranking Plans (WQRPs) some 25 years ago for all Bulgarian rivers. These plans were devised according to definite criteria, such as water use requirements, ecological requirements (taking into account the self-purification capacity of the reach), economical and social factors, as well as the available (and affordable) wastewater treatment technologies at that time. These same plans, slightly modified and updated, are still used by the MEW and RIEWs today.

The water quality status of the Struma (Strimon), Mesta (Nestos), Arda, Maritza (Evros, Merich) and Tundja Rivers has been assessed up to the summer of 1998, and show the river reaches ranks and pollutants loads.

The current water quality status of the rivers under consideration has been assessed according to NIMH water quantity for river discharges of 95% probability for minimum mean monthly values, calculated for the following points: 5 at Struma, 2 at Arda and 4 at Tundja. Another source of data is the NCESD, concerning the relevant pollutant concentrations, measured from 1993 to 1994 at the same points where the river discharges were measured by the NIMH. These are very close to those calculated with 95% probability for the minimum mean monthly values. This allows the comparison of the pollutants loads calculated on the base of the information described above with the "standard loads", calculated on the base of the relevant standard rivers reaches ranks.

As can be seen in the figures above, the measured pollutant concentrations in the rivers and the associated loads are in general below the standard limit values for the relevant reaches and ranks, according to the adopted WQRPs. The only exception is the Maritza River, where the values of some parameters exceed the standards. The favourable current ecological status of most of the rivers is mainly a result of the drastic (more than 10 times) reduction in industrial activities and the present stagnation of the economy. Industry in the Maritza River basin however is still discharging large quantities of untreated wastewater, causing ecological standards violation. Since there are signs of economical growth in Bulgaria, a gradual increase in the pollution load on all rivers has to be expected in the near future.

6. Conclusion

Special attention should be paid to the further development of the national water quantity/quality monitoring system, which need to be reorganised and equipped with modern equipment, including contemporary information systems for data acquisition, transfer, processing and dissemination. Since the available water quality monitoring and data acquisition and processing systems are not sufficiently reliable, research and feasibility studies are needed to determine exactly what future investments is needed in ecological issues concerning the river basins of the Struma, Arda, Tundja, Mesta and Maritza Rivers. Since all these rivers are transboundary and belong to the Aegean Sea basin, the importance of such investment is obvious and of great international concern.

CROATIA

J. MARGETA
University of Split
Split, Croatia

1. Introduction

Rivers and seas form the largest part of the international borders of Croatia, primarily with Hungary, Yugoslavia and Bosnia-Herzegovina (Figure 1). Croatia receives water from Slovenia, Hungary and Bosnia and Herzegovina (B&H), while Yugoslavia receives all river flows from the parts of Croatia that are in the Black Sea catchment area. The remaining waters in Croatia flow into the Adriatic Sea. The Croatian water supply is very much connected to that of its neighbouring countries with whom co-operation is needed to achieve rational and sustainable water resources management.

International obligations are guided by international treaties, legislation and government policy. These must seek to bridge legal and institutional gaps in water resources management among several countries. Croatia has signed international treaties on water resources management relations with the Republic of Hungary and Slovenia.

This text presents basic characteristics of these treaties with reference to water quality. It provides summaries of two basic laws that support water resources management in Croatia and transboundary waters with special reference to water quality. Information on current water monitoring activities is also presented.

2. Legal Framework for Water Resources Management

Water resources management in Croatia is conducted in accordance with the Water Act ("National Gazette" No. 107/95). It is framework legislation, which has generated a number of subsidiary laws on water resources management. Various provisions of the law are fundamental for understanding the water resources management system in Croatia, and are important for supporting transboundary water resources management policies. On the basis of the Water Act, the legal entity for water management, the Croatian Water Management Enterprise (CWME) (Hrvatske vode), was established.

J. Ganoulis et al. (eds.), Transboundary Water Resources in the Balkans, 107–116.

Figure 1 Rivers in Croatia.

The provisions of the Act refer to surface and underground continental waters including estuaries, mineral and thermal water with some exceptions, drinking water sources in territorial seas, and the protection of seas against pollution from land and islands. The director and secretary of the State Directorate for Water are responsible for maritime affairs and determining the boundary between continental and seawaters.

The Water Management Master Plan (WMMP) of the Republic of Croatia should be based on scientific investigations, continuous monitoring of distribution of states and phenomena connected with waters and their use. It should take into consideration overall environmental protection and the specific characteristics of water related problems of each river basin. The WMMP should also provide for the development and co-ordination of environmental protection, forest management on a state level and the interior navigation system. The WMMP should also be able to adapt to changes in the water system, and to economic and social development.

3. Strategy of Water Quality Protection in Croatia

A 1999 amendment to the Water Act provides for a Water Protection Master Plan (WPMP) to cover water quality management of territorial and international waters. This is of particular significance for transboundary water resources management. The WPMP requires investigations and analyses of water quality, categorisation of waters by quality, protection of water, measures for accidental and unexpected water pollution, and the construction of wastewater treatment plants and facilities. It also details sources and methods of financing, and a directory of parties and legal entities in charge of implementing the WPMP, their authority and responsibilities. The aim of the WPMP is to manage water on the principles of water system unity and sustainable development.

The WPMP rests on the principle of prevention that includes planning and measures required for water protection even when there is no firm evidence of a change in water quality. Preventive measures are activities for the prevention and restriction of discharge of detrimental or other matter that could cause pollution or contamination of water. Pollution control ensures the continuous monitoring of wastewater discharge at its source

The principle of using of the best technology available means saving raw material and energy, excluding dangerous matter from technological processes, and decreasing the quantities and noxiousness of matter before it is discharged. The "polluter pays" principle means that for every incidence of water pollution, a fine is collected according to the degree of pollution caused. The polluter is also obliged to reimburse any costs of cleaning and removal, as well as to recompense for any damage that is a direct consequence of the pollution he caused. Continuous exchange of information on water quality with boundary countries is important for informing the public and deciding on any water protection measures that need to be undertaken.

3.1 INVESTIGATIONS AND ANALYSES OF WATER QUALITY

The following investigations and analyses of water quality are carried out for surface water, ground water and coastal waters.

- Investigations and analyses of fresh water and sea water quality are carried out in order to determine water type i.e. to assess quality and to determine the cause of any change in quality, as well as to establish and apply measures required for water protection.

- Investigations and analyses of fresh and seawater are based on established monitoring programmes, which also detail how their implementation is to be financed.

- Programmes for transboundary water quality analyses and monitoring, which are the subject of treaties on water resources relations between the Republic of Croatia and neighbouring countries, are published in the "National Gazette" as annexes, and are components of the plans explained in Annex D-1.

- Results of water quality analyses of the above plans as well as the results of the National Programme of Monitoring within the Trans National Monitoring Network (TNMN) of the Danube catchment, are part of the programme of the Permanent Commission of the Convention of the Danube River Protection.

- Water quality monitoring programmes from these plans are the responsibility of the State Directorate for Water, and implemented by the Main Water Management Laboratory of the CWME (Hrvatske vode - Glavni vodnogospodarski laboratorij

- Water quality monitoring programmes of state waters (national monitoring programme), internal sea water and territorial sea polluted from the land, as well as programmes monitoring sources of their pollution on the land (Protocol on Mediterranean Sea Protection Against Pollution from the Land of the Convention on Mediterranean Sea Protection Against Pollution (LBS) programme) are established and implemented by the CWME, with the consent of the State Directorate for Water and the agreement of the State Directorate for Nature and Environment Protection in connection to the LBS - programme.

- The CWME is obliged to publish an annual report with the results of all fresh and seawater quality analysis programmes, as well as a report assessing change in water quality every five years.

Every five years the State Directorate for Water submits a report to the National Committee for Water, Government of the Republic of Croatia and the Croatian Parliament.

3.2 UNEXPECTED POLLUTION FROM OUTSIDE CROATIA

Incidental pollution refers to incidents, when due to low discharge or other circumstances, there is a possible hazard or decrease in water quality of the watercourse or other recipient into which the waste water is discharged. It may be caused by an unexpected spill of hazardous or other substances that can decrease the water quality or contaminate the surface and ground water or sea due to pollution from the land. The measures, procedures and operation plans to be used in such situations are regulated by law.

Measures to be undertaken in case of unexpected pollution are:

- To inform the public and the responsible authorities, and to implement the operation plans in order to control spreading and to eliminate the unexpected pollution.

- To determine the causes, responsible parties, type and scope of the pollution, and to assess the degree of endangerment to the ecological function of water, and to human health and life, and to gauge the possibility of pollution spreading.

- To keep the pollution from spreading, to inform the public and water users on water quality, and to prohibit water use if necessary.

- To mitigate the consequences of pollution in accordance with operation plans, and to prevent any further pollution from this source

The measures to be undertaken depend on the degree of endangerment to the particular water category.

3.3 PARTIES RESPONSIBLE FOR IMPLEMENTING THE MEASURES

The following parties or agencies are responsible for implementing the measures:

- State Directorate for Water.
- The CWME.
- Users whose activities may have an impact on the water quality of the rivers.
- Parties that perform municipal services.

4. International Treaties on Water Resources Management

Within the territory of the Republic of Croatia, there are numerous transboundary rivers of various sizes (see Annex D-2), the management of which has to be co-ordinated with neighbouring countries. An inventory of transboundary rivers in the Republic of Croatia is given in the following text, as well as water quality category that can be achieved by co-ordinated actions. Croatia has signed international treaties on transboundary water resources management with the Republic of Hungary and Slovenia. The Croatian Parliament enforces these treaties as part of the Strategic Plan for Water Quality Management. A summary of these treaties is provided below The Republic of Croatia has not yet signed any such treaties with B&H or Yugoslavia. Croatia is also an active participant in all activities connected with the Danube River and protection of the Mediterranean Sea (MAP).

112

Annexes

ANNEX 1A

Programme of water quality monitoring on transboundary waters based on the signed agreement on water resources management relations between the governments of the Republic of Croatia and the republic of Hungary.

Basis for the monitoring programme: Agreement on water resources management relations between the governments of the Republic of Croatia and the Republic of Hungary ("National Gazette - International Treaties" No. 10/94)

Water quality involves monitoring on the following rivers Dunav, Mura and Drava

Frequency of sampling on the rivers is also agreed upon and regulated by the treaty.

Indicators of water quality analyses on the Mura and Drava rivers within the programme are physical-chemical indicators, nutrients, metals and metalloids, organic compounds (pesticides), microbiological indicators and biological indicators.

On the Danube River, the programme includes all aforementioned indicators as well as radiological indicators.
Note that the Permanent Croatian-Hungarian Commission determines some water quality indicators for Water Resources Management.

The State Directorate for Water is responsible for programme implementation.

The Main Water Management Laboratory of the CWME in co-operation with other licensed laboratories with national and international experience performs water analyses.

ANNEX 1B

Programme of water quality monitoring on transboundary waters based on the signed agreement on water resources management relations between the governments of the Republic of Croatia and the Republic of Slovenia.

Basis for the monitoring programme: Agreement on water resources management relations between the governments of the Republic of Croatia and the Republic of Slovenia ("National Gazette - International Treaties" No. 10/94)

Water quality involves monitoring on the rivers Mura, Drava, Sava, Kupa, Dragonja, Sutla i Čabranka.

Frequency of sampling on the rivers is also agreed upon and regulated by the treaty.

Indicators of water quality analyses on the rivers Mura, Drava, Sava, Kupa for the sampling performed twelve times a year are: physical-chemical indicators, nutrients, metals and metalloids, organic compounds (pesticides), microbiological indicators (twice a year), biological indicators (twice a year), radioactivity (four times a year, for the river Sava only).

Indicators of water quality analyses on the rivers Sutla, Dragonja i Čabranka for the sampling performed four times a year are: physical-chemical indicators, nutrients, metals and metalloids, organic compounds (pesticides), microbiological indicators (twice a year), biological indicators (twice a year).

The State Directorate for Water is responsible for implementing the programme.

The Main Water Management Laboratory of the CWME in co-operation with other licensed laboratories with national and international experience performs water analyses.

In accordance with the above commitments the state budget provides annual funds for all monitoring activities as well as other activities connected with data exchange and consultation. The purpose of monitoring is to achieve the required water quality on certain parts of watercourses and sea i.e. boundary line.

ANNEX 2

This law also determines the categories (water quality) of Croatia's transboundary rivers. Details are given in Table 1.

114

TABLE 1.Categorisation of transboundary waters and seas in Croatia.

Natural recipients	Parts of the watercourse or sea for which the water category is determined	Water Category
VERY SENSITIVE AREAS		
	ground water which are used or planed to be used for water supply	I
	mountain springs to settlements	I
	watercourses to settlements in karst areas	I
	waters in national parks and parks of nature	I
TRANSBOUNDARY WATERS		
Dunav		II
Karašica (Baranja)		II
Karašica kanal (Baranja)		II
Borza		II
Hatvan		II
Travnik		II
Drava		II
Dombo		II
Ždalica		II
Izidorijus		II
Zajza		II
Mura		II
Zelena		II
Pritok Trnave (Dravske)		II
Sava	boundary with Slovenia to Zagreb	II
	from Zagreb to Sisak	III
	from Sisak to the boundary with Yugoslavia	II
Bosut	to the mouth in Bid	II
	from the mouth in Bid to the boundary with Yugoslavia	III
Studva		II
Smogva		II
Ilinački-Boris-Granični Ilinci-Šidski		II
Lipa		II
Una	spring Martin Brod	I
	boundary with B-H - mouth into Sava	II
Čabranka	spring - Čabar	I

Natural recipients	Parts of the watercourse or sea for which the water category is determined	Water Category
	Čabar to the mouth	II
Kupa	spring - Metlika	I
	Metlika - mouth of Korana	II
Glina	spring Topusko	I
	Topusko - mouuth into Kupa	II
Glinica		II
Mađarac		II
Podzvidska		II
Grabarska with tributary		II
Šiljkovača with tributary		II
Vidovska		II
Velika Jaza		II
Korana	from Plitvička jezera to Slunj	I
	from Slunj to Kupe	II
Bregana	from spring to Bregana	I
	from Bregana to its mouth into Sava	II
Sutla	from spring to Klanjec	I
	from Klanjec to the mouth into Sava	II
Dragonja		II
Butišnica		II
Mračaj		II
Neretva	from boundary with Bosnia-Herzegovina to the sea	II
Konavoštica		II
System "Baćinska jezera – Rastok"		II
System "Trebižat - Vrljika - Ričica"		II
SEA		
Sea in the area of disposal of water Area outside the circle around the outfall's from the land diffuser, of 300 m dia.		II*
More in specially protected and sensitive areas		I*

* Refers to seawater categorisation prepared by the State Directorate for Nature and Environment Protection together with the Ministry of Maritime Affairs, Transport and Communications in co-operation with the State Directorate for Water on December 31st 1999. Water categorisation will be published in the "National Gazette" as an amendment to Annex D-2.

5. Conclusion

The management of national and international water resources and seas in Croatia is entirely regulated by law. The laws are for the most part adjusted to EU legislation, of particular strategic interest to Croatia. All legislation for optimal water resources management is already in force and only a few supporting acts are missing. Croatia has signed all relevant international treaties and agreements that refer to terrestrial waters and the Mediterranean Sea. It has already been agreed that the management of transboundary water resources should be co-ordinated with the neighbouring countries of Hungary and Slovenia, in accordance with the commitments that are part of Danube River agreements. The monitoring programme and all other activities connected with regular and incidental situations are carried out in accordance with these treaties. Croatia is expected to sign treaties on water resources relations with B&H and, as well as with Italy with regard to Adriatic Sea management.

6. References

1. Water Act, National Gazette No. 107, 1995.
2. National Water Protection Master Plan, National Gazette, No. 26, 1999.

FYR OF MACEDONIA

B. MICEVSKI, O.AVRAMOSKI
Institute of Biology, Faculty of Natural Sciences and Mathematics
Skopje, FYROM

1. Introduction

Of the five watershed areas in the Former Yugoslav Republic of Macedonia (FYROM), the watershed areas of the rivers Vardar/Axios, Crn Drim and Strumica account for 99.33% of the total territory. The watershed of the Vardar/Axios River covers about 80% of the total territory of FYROM, and that of the Crni Drim and Strumica rivers about 13% and 7% respectively. The other three main natural water bodies in FYROM are Lakes Ohrid, Prespa and Dojran, and these are water resources of a transboundary nature, see Figure 1.

2. Institutional Arrangements

According to state legislation, the management of water resources in FYROM is divided among the Ministries of Agriculture, Forestry and Water Economy (MAFWE), Urban Planning, Construction and Environment (MUPCE), Health (MPH) and Economy (ME).

MAFWE is responsible for monitoring and examining water resources, maintenance and improvement of the water regime, exploration and use of the water resources, water pollution protection and so forth. The Republic Office of Water Resources Management at MAFWE may, upon request, issue a water permit, which defines a site-specific effluent norm based on ambient standards in protecting the receiving water body.

MUPCE and its Republic Department for Protection of the Environment are, among others, responsible for the protection of natural resources. The environmental department of MUPCE compiles a Register of Polluters based on data provided from the following agencies:

- Republic Hydro-meteorological Institute, Skopje.
- Republic Health Care Institute, Skopje, and its subsidiaries of municipal health care institutes.

117

J. Ganoulis et al. (eds.), Transboundary Water Resources in the Balkans, 117–124.
© 2000 *Kluwer Academic Publishers.*

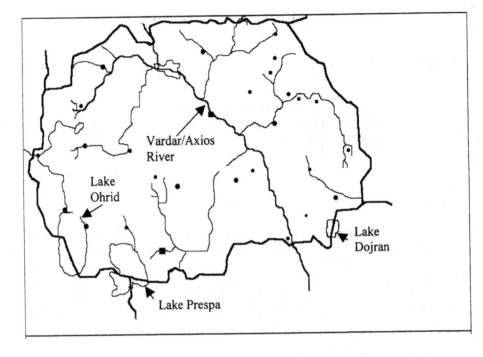

Figure 1. Rivers and lakes in FYROM.

- Occupational Health Institute, Skopje.
- Preventive Medical Care Institute at the Military Hospital, Skopje.
- Mining Institute, Skopje.
- Veterinary Institute, Skopje.
- Centre for Application of Radioisotopes in Industry, Skopje.
- Hydro-biological Institute, Ohrid.
- Construction Institute "Macedonia", Skopje.
- Institute for Earthquake Engineering and Engineer Seismology, Skopje.
- Central Institute for Occupational Safety, Fires and Environmental Protection. Skopje.

However, the Register of Polluters is only occasionally updated, and the fact that little analysis is done makes policy decision making difficult. Various environmental data different to those collected by MUPCE are also collected and statistically processed by the Statistical Office of FYROM. However despite these efforts, only partial data are available related to environmental issues.

Exploitation of water resources in electricity power production is the responsibility of the Ministry of Economy. The Ministry of Health monitors the quality of drinking water and the water used in the food industry.

TABLE 1. Ministries responsible for monitoring the water quality and protection.

Ministry	Address	Contact person
Ministry of Urban Planning, Construction and Environment	Dame Gruev 14 366-930	Aleksandar Nastov
Ministry of Agriculture, Forestry and Water Economy	Vasil Gjorgov b.b. 113-045, 211-997	Aleksandar Trendafilov
Ministry of Health	50. Divizija b.b. 113-429, 113-014	Dr Elisaveta Stikova
Ministry of Economy	Samoilova 10 119-628, 111-541	Ljupcho Trajkovski

3. Legislation

Several provisions address the protection of water resources:

- The Law on Water Regime of Transrepublic and Transboundary Waters (*Official Gazette of SFRJ. No. 2/74 and 24/76*) and in detail under the Regulation on Classification of the Waters of Transrepublic Streams, Transboundary Waters and Coastal Waters of the Sea of Yugoslavia (*Official Gazette of SFRJ. No.8/78*).

- The Law on Protection of Lakes Ohrid, Prespa and Dojran (*Official Gazette of RM, No.45/77 and improvements 8/80, 51/88, 10/90, 62/93*) and its Programme for Protection of these lakes (*Official Gazette of SRM No. 7/87*).

- The Law on Protection of Waters (*Official Gazette of SRM No. 6/81 and improvements 13/82, 37/87, 15/88, 20/90, 23/90, 24/91 and 83/92*) and in detail under the Regulation on Water Classification (*Official Gazette of SRM No. 4/84*), Regulation on Categorisation of Streams and Lakes (*Official Gazette of SRM No. 9/84*), Regulation on Maintenance of the Protected Zones Around Drinking Water Sources (*Official Gazette of SRM No. 17/83*).

- The Law on Environment and Nature Protection (*Official Gazette 69/96*).

4. The Monitoring of Water Resources Quality

4.1 RESPONSIBLE INSTITUTIONS

According to the Law for Protection of Waters (*Official Gazette of SRM No. 6/81*), the Republic Hydro-meteorological Institute in Skopje (RHMI) is the institution responsible for monitoring the qualitative and quantitative characteristics of waters in FYROM. The RHMI addresses the following issues:

- Estimation of the extent of pollution and determination of pollution trends, to assist the programme protecting water from pollution and exploitation.

- Collection of data for devising and implementing a detailed programme for water resources.

- Preparation of information and studies on changes in water quality along control points of the rivers of different watersheds.

- Preparation of information on the nature and development of sources of pollution strongly affecting water quality.

The quality of surface waters from a public health viewpoint is monitored by the Republic Health Care Institute and by 10 Departments of Public Health throughout the country, as well as by the Preventive Medical Care Institute at the Military Hospital in Skopje. Various research institutes also collect environmental data during their research and scientific work.

4.2 THE MONITORING PROGRAMME

The current programme for monitoring the quality of surface waters in FYROM consists of two monitoring schemes. The monitoring network for transboundary waters includes 12 monitoring points. The so called national monitoring network covers the more important rivers, and natural and artificial lakes in the republic at 60 sampling points (taking samples in March, May, June, August, September and November), and at 13 sites (taking samples 12 times a year.

TABLE 2. Parameters measured at each sampling site by the RHMI.

Parameter	Description
Hydrological parameters	Measuring of the water level and quantity.
Physical and organoleptic features of the water	Temperature of water and air, visually detected waste materials, odour, colour, turbidity, pH, conductivity.
Oxygen status	Dissolved oxygen, saturation with oxygen, BOD_5, COD by dichromate and by potassium permanganate.
Mineralisation	Residue (total, volatile and fixed), dissolved matter (total, volatile and fixed), suspended matter (total, volatile and fixed).
Cations and anions	Bicarbonate, carbonate, hydroxide, sulfide, chloride, potassium, magnesium, sodium, calcium, hardness.
Biogenic parameters	Ammonia, nitrates and nitrites, phosphorous.
Toxic matters	Cyanides, phenol, anionic surfactants, sulfides, iron, manganese, lead, zinc, cadmium, chromium-hexavalent and, if necessary, cooper, nickel and cobalt and others.
Microbiological parameters	Aerobic bacteria at 22°C and 37°C, MPN of coliform bacteria, isolation and identification of E.coli and Salmonella.
Saprobiological parameters	Productivity (for natural and artificial lakes only), measured at 25 sites three times per year (spring, summer and autumn).
Radiological parameters	Total beta radioactivity measured at 8 points 6 times per year.

4.3 CLASSIFICATION OF THE WATERS IN FYROM

The Classification of Waters classifies the surface water (streams, and natural and artificial water bodies) and ground water into four classes of water quality:

Class I: water can be used for water supply and in the food industry and, if surface water for fish breeding (e.g. Salmonidae).

Class II: water can be used for recreational purposes and water sports, fish breeding (Cyprinidae) and, if previously treated by coagulation, filtration, dosification and other ordinary methods, for water supplying and in the food industry.

Class III: water can be used for irrigation or industry (but not the food industry) if previously treated by coagulation, filtration, dosification and other ordinary methods.

Class IV: water can be used for other purposes if previously properly treated.

Classification of the surface waters during the year 1994, covered by the monitoring programme implemented by the RHMI and according to the appropriate parameters is presented in Table 3.

TABLE 3. Evaluation of surface water quality in FYROM.

Sampling site	Quality regulated by the Law	Quality evaluated in 1994
River Vardar/Axios		
01. v. Vrutok	1	1-2
02. v. Balin Dol	2	1-2
03. v. Sarakinci	2	3-4
04 . v. Jegunovce	2	3-4
05. Skopje-Saraj	2	3-4
06. Skopje-Vlae	2	2-3
07. Bank Complex	2	3
08. v. Jurumleri	2	4
09. v. Basino Selo	2	4
10. Veles	3	4
11. Post. r. Babuna	3	4
12. v. Nogaevci	3	3-4
13. Staro Gradsko	3	3-4
14. Demir Kapija	2	3-4
15. Gevgelija	2	3-4
River Bistrica		
16. v. Jegunovce mouth	2	4-out
River Lepenec		
17. Asphalt Base	2	3-4
18. Skopje mouth	2	4-3
River Treska		
19. v. Izvor	1	1-2
20. v. Bigor Dolenci	2	2-3
21. Skopje-Saraj	2	2
River Zajaska		
22. Kicevo	3	2-3
River Pcinja		
23. v. Pelince	2	2
24. Katlanovska Banja	2	3-2
River Kriva Reka		
25. v. Klecovce	2	2-3
River Kumanovska		
26. Lipkovsko Lake-dam	1	2-1
27. v. dobrosane	3	out
River Bregalnica		
28. v. Trabotiviste	2	3
29. v. Oci Pale	2	3-4
30. v. Istibanje	2	2-3

TABLE 3. Evaluation of surface water quality in FYROM (Cont.)

31. v. Krupiste	3	3-4
32. Post Stip	3	4-out
33. v. Ubogo	3	4-3
River Kamenicka		
34. Mak. Kamenica	3	2-3
River Zletovska		
35. Zletovska mouth	3	2-3
River Crna Reka		
36. v. Topolcani	3	3
37. v. Novaci	3	4
38. v. skocivir	3	4-out
39. Tikves Lake-dam	2	2
40. v. Palikura	2	2
River Dragor		
41. after Bitola	3	out-4
River Eleska		
42. v. Brod	2	2-3
Dojran Lake		
43. Nov Dojran	2	4
44. Star dojran	2	4-3
River Strumica		
45. after Radovis	3	4-out
46. v. Novo Selo	3	3-4
River Crni Drim		
47. after Struga	2	2
48. Debar Lake-dam	2	2
49. after HE Spilje	2	2
River Radika		
50. v. Zirovnica	1	1-2
51. Boskov Most	1	2
52. Mavrovo Lake-Anovi	2	2
Prespa Lake		
53. v. Otesevo	1	2
54. v. Pretor	1	2
55. Post Resen	2	4-out
Ohrid Lake		
56. St. Naum	1	1-2
57. Metropol Hotel	1	1-2

TABLE 3. Evaluation of surface water quality in FYROM (Cont.)

58. Ohrid port	1	2-3
59. Ohrid -city beach	1	3
60. Biser Hotel	1	1-2

5. Enforcement

There are many agencies monitoring environmental parameters, however they are not directly responsible for enforcement, and the co-ordination between these agencies is inadequate. The MUPCE responsible for quality control and enforcement does not have enough staff to fully carry out its functions. Most of the inspectors in MUPCE do not have an environmental focus since their first responsibility is to the departments of planning and construction.

The current regulations need revision of the standards for water quality, particularly since some of the currently listed substances are already forbidden. The emission standards for wastewater are stringent and considered unrealistic. This makes them difficult to enforce. The Law on Waters, for example, concentrates mainly on the economy of waters and not on environmental issues. Of the 210 provisions in the law, only 9 refer to water resources protection.

The Law on Environment and Nature Protection and Promotion has addressed some of the key weaknesses of the current legal system. The National Environmental Action Plan (NEAP) suggests establishing a Ministry of the Environment with an Environment Institute (as a corporate body of the Ministry), an Inspection Office and a Department for International Co-operation and Training. Also planned are a transboundary pollution programme and information system for environment quality monitoring, and a link to international databases and to the European Agency of Environment

6. References

1. Gashevski, M.,1972. Vodite na SR Makedonija. ZID Nova Makedonija,Skopje.
2. Nastov, A.,1998. Upravuvanje so zashtitata na vodite vo Republika Makedonija. Vodostopanstvo na Makedonija, 1.
3. Sibinovik, M.,1987. Ezerata na Makedonija. Prespansko i Ohridsko. Zavod za vodostopanstvo na SRM.
4. Statistichki godishnik, 1994. Zavod za statistika na Makedonija.
5. Statistichki godishnik, 1996. Zavod za statistika na Makedonija.
6. Statistichki godishnik, 1997. Zavod za statistika na Makedonija.

GREECE

E. SIDIROPOULOS, D. TOLIKAS
P. TOLIKAS
Faculty of Engineering
Aristotle University of Thessaloniki
Thessaloniki, Greece

1. Introduction

International co-operation is an inevitable aspect of modern scientific and technical practice. It has nowadays been definitely realised that national borders cannot contain the effects of pollution and that cross-border water resources management cannot be restricted within the confines of a single state.

Surface cross-border water resources are lakes and rivers separated by the borders of two or more neighbouring countries. In the same way cross-border groundwater resources are aquifers underlying such borders. The main issues concerning joint cross-border water resources management are classified and discussed. Co-operation with neighbouring countries is encouraged in the field of Hydroinformatics and in a variety of common projects. The main general issues are outlined and co-ordination measures are proposed.

The particular problems of Greece pertaining to water management with a special reference to cross-border resources are summarised. The factors influencing the overall situation are distinguished into geomorphological, climatological and hydrological on the one hand and socio-economic and administrative on the other.

Legislation aiming at a rationalisation of water management has been introduced in Greece, but still meets obstacles in its full implementation. An effort toward creating a national hydrological data bank is noted. The data bank is in line with the spirit of the new legislation and will facilitate integration into appropriate European Union (EU) information systems.

2. Water Resources Data and Issues in Greece

According to internationally established criteria, Greece is a privileged country with regard to fresh water deposits. Indeed, according to the United Nations (UN) classification, countries are considered almost dry when their annual per capita renewable fresh water deposits lie between 2000-1000 m^3/person, while they are considered to be completely dry when this indicator falls below 1000 m^3/person [5]. Based on data of the last decade, Greece presents a value of 5828 m^3/person, having

J. Ganoulis et al. (eds.), Transboundary Water Resources in the Balkans, 125–135.
© 2000 *Kluwer Academic Publishers.*

thus a significant advantage over most European countries. It follows that Greece should not have a water availability problem. However, a number of adverse factors contribute to serious water problems. These factors have been classified and discussed by Mylopoulos [4,5] and are summarised here. The following broad categories are distinguished:

2.1 GEOMORPHOLOGICAL AND HYDROLOGICAL FACTORS

These factors include:

- Unequal temporal and spatial distribution of water resources
- The majority of precipitation occurs during the winter, while the actual needs are highest during the summer. Also, there is an unequal water resource distribution between the moist mountainous regions and the dry coastal areas and islands. Especially, the climatic regime of the Aegean islands with intense winter rains and summer droughts has posed particular challenges to water management [3].
- Absence of large surface systems.
- This factor has led people to use groundwater resources and to consequent over-exploitation. Five large river systems can be distinguished, four of which are transboundary.
- Saltwater intrusion into coastal aquifers
- This factor is combined with over-exploitation.
- Extreme drought conditions of recent years.

2.2 SOCIOECONOMIC FACTORS

These factors include:

- Predominant agricultural use of water.
- Indeed, 85% of total consumption serves agricultural activities, while industrial activities absorb only about 8%, compared with corresponding mean international values of the order of 20%. Agricultural consumption takes place mostly in the summer and this fact combines unfavourably with the unequal temporal distribution of water resources.
- Rapid urbanisation
- Nearly half of the country's population lives within the two large urban complexes of Athens and Thessaloniki, thus creating huge needs of water supply and of appropriate networks and accompanying hydraulic works.
- Orientation toward supply management.
- Supply rather than demand management has been the dominant managerial approach until today.

- Under-pricing of water.
- In the eyes of the public, water is considered to be a free public good, making it difficult for the government to apply various economic policies or measures.
- Overall difficult economic conditions
- Inflation rate around 10% in recent years, increasing unemployment and various other economic restraints have driven rapid economic development to the top of the list of national priorities, thus creating a rather negative climate for a systematic integration of environment and economy.

2.3 ADMINISTRATIVE AND LEGISLATIVE FACTORS

These include:

- Centralised administrative system.
- Sectored structure of water- related services.
- The multitude of water resources agencies has resulted in fragmented managerial practices. The jurisdiction of the various water authorities was determined according to water use criteria and not according to a hydrological based scheme.
- Water Policies

3. Legal Framework

In order to place water resources management into a more integrated and rational framework, a new law was enacted in 1987. The features and the significance of the new legislation are outlined by Mylopoulos [5]. Some of its aspects are considered here.

Under the new law, the general water resources management falls under the jurisdiction of the Ministry of Development. This ministry is responsible for general policy and co-ordination and it only transfers authority to other ministries over specific matters of water use, e.g. to the Ministry of Agriculture which is responsible for agricultural use, to the Ministry of the Interior which is responsible for water supply, etc. Also, the country was divided into 14 water districts, each one of which included one or more drainage basins. Also the law assigned a water management service to each water district. Again all these services function under the co-ordination of the Ministry of Development. Each district is held responsible for carrying out water balances and for that purpose provision was made for an archive of hydrological data.

For a variety of reasons analysed by Mylopoulos [5], the new legislation was never fully implemented.

3.1 NATIONAL DATA BANK

As a response to the call for archives of hydrological data, a database system was designed for the management of Greece's hydrological, hydrogeological and meteorological information on a nation-wide basis. In order to carry out this task a number of participating organisations co-operated within the framework of the HYDROSCOPE project, which was funded by the EU. The salient features of this project are described by Tolikas et al.[8]. The organisation of information pertaining to groundwater hydrology are discussed by Georgiadis et al. [1, 2], while treatment of hydrometeorological data is presented by Sakellariou et al [6] and by Tsakalias and Koutsoyiannis [9].

The HYDROSCOPE database system's structure respects the directives of the new legislation. Namely, it is based on the geographical division of the 14 water districts and it classifies the data according to purely hydrological criteria and not according to water use or according to data origin. Although the data classification is uniform, the data base system does not follow the old-fashioned strictly centralised system, but presents a distributed structure.

The various water-related authorities will by necessity be using the new database system and it is hoped that this practice will facilitate and promote a fuller implementation of the new legal framework.

At this stage, the majority of the hydrological data have been introduced into the database system. However, at present, the database is not yet publicly available. As soon as it becomes available, hydrological assessments of various kinds will be feasible. Also, integration into appropriate European Information Systems will be possible, as well as co-operation with neighbouring countries.

3.2 MONITORING PROGRAMMES

The Ministry of the Environment, Physical Planning and Public Works (MEV) has organised the National Network for monitoring the quality of the country's surface waters. According to MEV sources (official MEV Webpage), samples are taken on a monthly basis and are analysed in regional laboratories, belonging to the Ministry's Network of Laboratories. Measurement data are archived into a special databank and they are assessed for the purpose of locating environmental pressures and taking the appropriate measures. Figure 1 shows the distribution of the network stations.

○   Sampling point

Figure 1. Distribution of network stations in Greece.

The parameters measured are the ones prescribed by the respective EU Directives according to water use (water for drinking, water for swimming). In parallel with the above National Network, automatic measuring stations are installed for continuous water quality monitoring of the transboundary rivers Evros, Nestos, Strymon and Axios at their entry points into Greece. The continuously measured parameters are pH, temperature, conductivity, dissolved oxygen, opaqueness and elevation. Through telemetry, measurements are automatically recorded and signals are transmitted, in case any of the measured parameters exceed predefined limits.

The international Helsinki treaty (March 1992) for the protection of transboundary resources has been ratified (October 1996). In the context of this treaty, co-operation in monitoring transboundary resources is promoted with Bulgaria and FYROM.

4. Transboundary water resources.

Cross-border water resources represent almost 25% of total natural water resources in Greece. The mean annual flow of the most important Axios/Vardar river (Greek-FYROM) was more than 5.000×10^6 m^3 in the 60's and is currently about 2.500×10^6 m^3. The Strimon, Nestos and Evros river basins (Greek - Bulgarian) are shown in Gigure 2. The mean annual flow of Strimon River is about 3.500×10^6 m^3

The main issues concerning joint cross-border water resources management are quantitative and qualitative. In both cases the ecological aspect cannot be overemphasised. These types of issues are interconnected, as the quantity of the water affects its quality (natural purification) and hence the ecosystem. The separation is suggested only for the sake of a methodological approach to the subject. The discussion of this section on transboundary issues and policies was presented at the 1999 HELECO Conference [7].

4.1 QUANTITATIVE ISSUES

The main issues regarding the quantity of the water with respect to the following are:

Environmental role of rivers

The flow rate of a river must be sufficient for it to perform its environmental role along its entire length from source to estuary. This may constitute the minimum commitment of all the involved countries in order to preserve ecosystems, national parks and reserves.

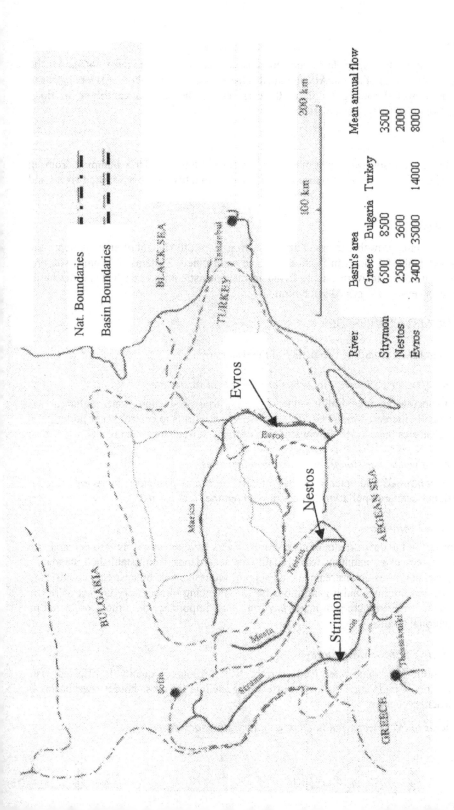

Figure 2. Strimon/Struma, Nestos/Mesta and Evros/Maritsa river basins.

Dams

The operation of existing dams and the construction of new ones must not ignore the environmental role of rivers. Already the discharge of some cross-border rivers has been decreased by half during the last 30 years. Dams should also contribute to flood mitigation of the downstream areas.

Pumping

Surface water resources and groundwater are interconnected. Thus pumping from an aquifer, usually for irrigation purposes, affects the quantity of the associated rivers and lakes.

Flood mitigation

The maximum allowable released discharge from dams, the height of embankments and the relief tanks for flood mitigation must be determined. Systems of prompt warning must be installed to warn people living in the downstream areas when the discharge exceeds a certain predefined safety limit.

4.2 QUALITATIVE ISSUES

The main issues regarding the quality of the water are:

Control of water pollution from urban and industrial wastewater.

Due to increasing urbanisation and rising industrial production, water pollution has dramatically increased in the last decades. However, efforts to control water pollution in some countries have led to the construction of many sewage treatment plants.

Control of water pollution from agricultural activities.

The uncontrolled and intensified use of fertilisers and pesticides in recent decades constitutes scattered pollution sources throughout the river basins.

Rights and obligations.

Countries on the downstream of cross-border rivers have the moral right to demand that upstream countries maintain water quality in accordance with established standards. Likewise, countries downstream should support requests made by upstream countries to national and international organisations for the financing of necessary projects. Priority is given to water resources of major environmental importance or to those protected by international treaties.

Radioactive or toxic waste storage.

When polluted groundwater feeds surface water resources quality is affected. The prohibition of radioactive or toxic waste storage within the cross-border river basins is of particular concern.

Measures to avoid water pollution as an aftermath of accidents.

4.3 RECOMMENDED CO-ORDINATED ACTIONS

The following actions should be undertaken to improve transboundary water quality:

Establish a committee of experts.
Every country should appoint engineers to participate in a committee of experts, who will act as consultants on technical issues to the competent authorities of their countries.

Establishment of a co-operative network
All countries' committee members will work together continuously through an Internet network. They will also participate in workshops organised at least every year. Thus, they will maintain a co-operative network in the southeast part of Europe for cross-border water resources.

The objectives of the co-operative network can be divided into the following categories

- Establishment of a network infrastructure.
- Design and installation of electronic networks.
- Development of a common databank for hydrological data.
- Development of common water balances and of mathematical models.

Co-operative tasks include:

- Joint management of cross-border surface and groundwater resources.
- Environmental protection of rivers along their entire length from source to estuary.
- Technical advice on prospective regional cross-border water resources planning.

Possible joint actions and measures:

- Allocation of water quantities.
- Determination of quality standards.
- Control of maximum discharges to be released.
- Control of quantitative and qualitative parameters.

The above list of tasks constitutes only an indicative set of activities, with which the network of experts could concern themselves. Existing problems in the Balkan area cannot be eliminated within a short period of time, and thus a water resources centre that would act as a co-ordinating body should not be seen as a short term solution.

Hydrodiplomacy is a fascinating new term that has come into existence shortly after the introduction of another successful analogous coinage, that of hydroinformatics. This is a happy coincidence. Indeed, hydrodiplomacy must be based on hydroinformatics. Moreover, the cultivation and practice of hydroinformatics will facilitate the hydrodiplomatic effort. This will be achieved through an inductive rather than deductive process. Instead of establishing regulation and control in a vertical, top to bottom fashion, the goal of overall harmonic co-operation will be approached in the Balkans through many particular, individual projects among interested neighbouring countries. Leaving sensitive issues aside and concentrating on non-problematic areas will keep the co-operative spirit alive.

5. Summary and Conclusions

An overall picture has been given of the main aspects of water resources management in Greece, with particular reference to transboundary water resources. Comments have been made on the main factors and issues both from a hydrological and from a socio-economic point of view. The current legislation has been outlined and its implementation discussed.

The compilation of a national data bank to encompass the totality of the hydrological, hydrogeological and meteorological information of the country is nearing completion. Data validation and data bank connection with mathematical models is planned for the immediate future. For these reasons data bank products and outputs are not yet publicly available. Full activation of the national database system will make hydrological assessments and water resources information reports more readily obtainable and they will include a greater variety of information. Specific information is currently being gathered from diverse sources and services, in support of the issues, views and estimates discussed in this report. Reference should be made to the National Monitoring Programme of the Ministry of the Environment and to plans for co-operation with our neighbouring countries.

6. References

1. Georgiadis, N., Ladas, S., Sidiropoulos, E. and Tolikas, P. (1993). Organising information related to Groundwater Hydrology. *HYDROSOFT 94*, Proceedings of a Conference, K. Katsifarakis (ed.), Computational Mechanics Publications.

2. Georgiadis, N., Ladas, S., Sidiropoulos, E. and Tolikas, P. (1995). Management of hydrogeological information. *Water Resources Management under Drought or Water Shortage Conditions*, Proceedings of the EWRA 95 Symposium, N. Tsiourtis (ed.) Nicosia, Cyprus. Balkema Publishing.

3. Glezos, M., Iakovides, I., Theodossiou, N. and Sidiropoulos, E. (1999) When every drop of water counts: Water management in dry islands. *Groundwater Pollution Control*, K. Katsifarakis (ed.) WIT Press.

4. Mylopoulos, Y. (1996) Sustainable Water Management in Greece. A Dream or a Vision?, *Proc. Int. Conference: Rational and Sustainable Development of Water Resources*, Canadian Water Resources Association, Collection Environnement de l' Universite de Montreal, No 6, Vol. II, pp 652 - 660.

5. Mylopoulos, Y. (1997) Legal framework and water policy in Greece. (in Greek) *Proceedings of a Conference on water resources management in the basin of Kozani, Ptolemaida and Amyntaion*, Technical Chamber of Greece, Section of West Macedonia.

6. Sakellariou, A., Koutsoyiannis, D. and Tolikas, D. (1994) HYDROSCOPE: experience for a distributed database system for hydrometeorological data. *HYDROSOFT 94*, Proceedings of a Conference, K. Katsifarakis (ed.), Computational Mechanics Publications.

7. Tolikas, D. (1999) Transboundary water resources. *Proceedings of the HELECO 99 Conference*, Technical Chamber of Greece, Section of Central Macedonia.

8. Tolikas, D., Koutsoyannis, D. and Xanthopoulos, T. (1993) HYDROSCOPE: Un systeme pour l' etude des phenomenes hydroclimatiques en Grece. *6eme Colloque International de Climatologie*, Proceedings of a Conference, Thessaloniki.

9. Tsakalias, G. and Koutsoyiannis, D. (1994) Hydrologic data management using RDBMS with differential – linear data storage. *HYDROSOFT 94*, Proceedings of a Conference, K. Katsifarakis (ed.), Computational Mechanics Publications.

ROMANIA

P. SERBAN
Romanian Waters Authority "Apele Romane"
Bucharest, Romania

R. M. DAMIAN
Technical University of Civil Engineering Bucharest,
Bucharest, Romania

1. The Concept of Integrated Water Monitoring

The most important activity in water management is the process of decision-making at different levels, namely local, regional, basin and national. Decisions concern the qualitative and quantitative water supply for various uses, flood control and the prevention and diminishing of the effects of pollution. Decisions are based on real-time information on characteristic parameters related to the aquatic environment, water use, and the state and operation of hydraulic structures. Data are obtained from an integrated water monitoring system, which is used for the integrated qualitative and quantitative management of surface waters and groundwater at river basin level. Romanian rivers are shown in Figure 1.

From the point of view of information, the concept of sustainable management of water resources implies firstly the creation of systems for integrated monitoring on each environmental, such as water, forest, soil and air, and secondly, the realisation of the environmental global monitoring system. This will include a selection of the information obtained for each environmental factor and is needed to determine the causal links between these factors.

Water monitoring also includes standardised, permanent long-term observations and measurements on selected characteristic parameters of water, to provide data for the water management process, to evaluate the state of the aquatic environment and to provide information on any trends.

An integrated water monitoring system should provide information on:
- Precipitation, runoff, water resources various uses of water.
- Water quantity and quality.
- All the aquatic objects (rivers, lakes, groundwater and the sea) and the interactions between them.
  Information should be integrated for use at four levels:
- Local level (e.g. reservoir, water use).
- Regional level.

J. Ganoulis et al. (eds.), Transboundary Water Resources in the Balkans, 137–151.
© 2000 *Kluwer Academic Publishers.*

138

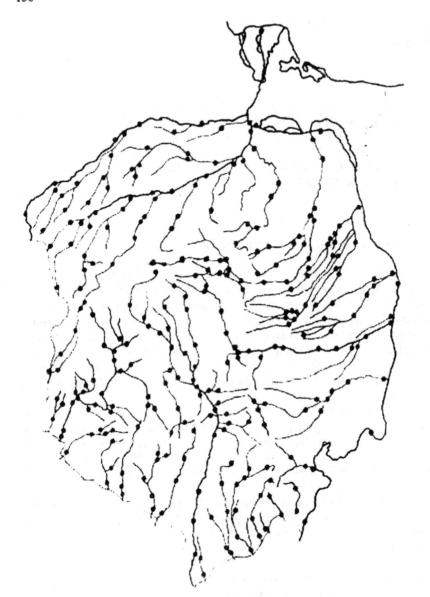

Figure 1. Rivers and monitoring stations (•) in Romania.

- Basin level.
- National level.

The purpose of integrated water monitoring is to provide data and information for:
- Up-dating knowledge on the state of the aquatic environment.
- Allocation of water resources to various categories of use.
- Warning and informing water users and the population in general in extreme or unexpected cases, such as high intensity rainfalls, floods, or accidental pollution, so that any adverse results may be diminished, and prevented in future.
- Checking whether users comply with the regulations for quality of wastewater sources and return flows.
- Determining the effects of human activities on the evolution of the aquatic environment and taking any necessary preventive or corrective measures.

2. Structure and Dynamics of the Water Monitoring Cycle

The monitoring process can be considered as a series of interrelated activities (Figure 2). The most important of these are deciding what information is necessary, defining the monitoring strategy, designing the network, sampling, laboratory analyses, data transmission, data analyses, data reporting and use of information in integrated water management.

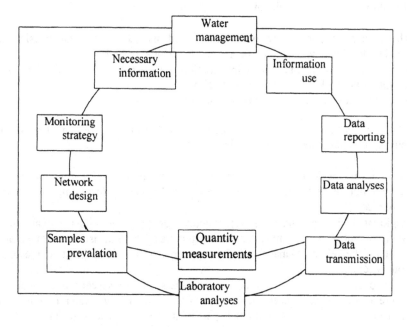

Figure 2. The structure of the water monitoring cycle.

In order to define the monitoring parameters one must firstly consider the need to obtain comprehensive and permanent information on the state and evolution of the aquatic environment, and secondly the kind of decisions that need to be taken in the field of quantitative and qualitative water management. Such decisions are of a tactical, strategic and operational nature.

Strategic and tactical decisions refer to establishing development directions in water management. Such directions include improving water management plans, establishing what works need to be undertaken to ensure an adequate water supply for various uses, and protecting water resources against exhaustion and pollution. These issues of water planning are included in the Master Plan for Water Management, which is updated every 5 years and submitted for approval to the Romanian Government. Strategic decisions also refer to Romania's participation at international conventions regarding the Danube and the Black Sea, as well as other important conventions and international programmes.

Operational decisions refer to:

• Optimal distribution of water resources to the various categories of use.
• Operation of the water management works.
• Combat against droughts and floods.
• Protection against accidental pollution.
• Safety of hydraulic structures.

Monitoring strategy should take into account:

• The means of performing observations and measurements, transmission and processing of data which could be automatic, manual or both.
• The means of dissemination of information, which may be by television, radio, computer network, radiotelephone, telephone etc.
• The levels at which the decisions are taken.

Water management decisions are generally taken at the following levels: local (objective), regional (county), basin and national.

• Local level decisions are taken by the managerial team of every objective on matters not of general importance, which only have a local impact on water resources.
• Regional decisions usually refer to matters such as reservoirs and water users, where operation has an impact on a restricted area. These decisions are taken at the level of the Water Management System, created for administrative reasons at county level.
• Basin decisions refer to all important matters such as reservoirs and water uses, where operation influences the quantitative and qualitative water regime of the entire river basin or of a significant part of it. These decisions are taken by each river basin branch of the Romanian Waters Authority (Apele Romane).

- National decisions concern the allocation of water originating from an area wider than a river basin, the use of water from the rivers which form Romania's state border, and for the operation of strategic reservoirs.
- Taking the type of decision into account, the monitoring system is organised into rapid and slow fluxes:
- Rapid flux refers to real time data that allow forecasts to be elaborated and the best operational decisions to be taken.
- Slow flux refers to the creation of a national database so that strategic decisions can be taken.

The **design of the water-monitoring network** should establish the measuring and sampling points, the elements that are to be measured and the frequency of their observation and measurement. The design should be based on the criteria specific to each monitored parameter (presented in the next chapter) and should take into account:

- Time and space variability of the monitored parameter.
- Human impact on the aquatic environment.
- Interdependency with other specific water parameters and environmental factors.

Sampling is preferably done automatically. Samples for most of the parameters are usually taken from both shores and from the water stream.

Sample analysis is performed in situ for the parameters that are more sensitive to modifications in environmental conditions, such as temperature, pH, dissolved oxygen and conductivity. Analysis for other parameters is usually done in the laboratory, because supplementary operations such as distillation or mineralisation are necessary, or because the analysis takes a long time (e.g. phenols, heavy metals, BOD5).

Data transmission is preferably done automatically, so that there is enough time for managers to take decisions regarding the prevention and limitation of possible negative effects.

Data analysis involves the comparison of data from different stations, the analysis of trends, the elaboration of the cause-effect relations, for example between water quality and pollution sources or land use and hydrological data.

Data reporting is done selectively, depending on the user (e.g. from the public or scientific sector).

In trying to achieve an integrated water monitoring system, one should consider the following paradox: the monitoring of a particular parameter for water requires knowledge of its time and space variability. This can be acquired only by observation and measurement. Consequently a water monitoring system is never static, but has its own dynamics in time and space in the form of a spiral (see Figure 3).

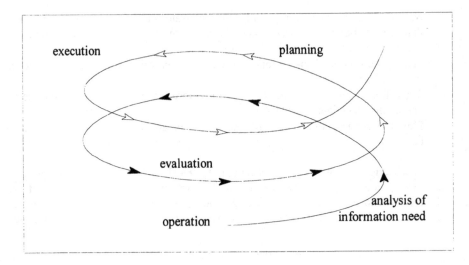

Figure 3. The dynamics of the water monitoring system.

3. Design Criteria for a Monitoring Network

The following are the design criteria for **rain gauge stations,** where measurements allow the elaboration of hydrological forecasts:

- 1 station per 150 km^2 in mountainous areas
- 1 station per 250 km^2 on plains.

The design criteria for **river hydrometric stations,** where water level and discharge are measured, are:

- Physico-geographical representativity for the basins with surface $100<F<500$ km^2
- Azonalities for the basins with surface $F<100$ km^2
- Significant changes of the hydrological regime.
- Linear interpolation for the middle and inferior sector of the river of $F>3,000$ km^2.

When deciding the position of river quality control stations, the following should be taken into account:

- The importance of the water stream at a national level, depending on its size (discharge, length) and uses.
- The homogeneity degree of water;
- The possibility of measuring the discharge associated with quality determinations;
- The existence of conditions affecting access and work.

The choice of which natural and artificial lakes are to be monitored depends on water uses e.g. drinking water supply, fish farming, irrigation, and electric power production.

The projection criteria for the well network where groundwater levels are measured are the following:

- 1 well/20 km^2 in areas where the piezometric level is close to the surface.
- 1 well/50 km^2 in areas where the piezometric level is very deep.

Water users should realise that it is their own responsibility to monitor water quality parameters for water intake and return flows .

The water authorities should take control measurements between 1 and 24 times a year, depending on the importance of the water use and its impact on the aquatic environment.

Water pollution sources are classified according to the following criteria, as presented in Annex 1:

- Dilution factor.
- Effluent quality.
- Magnitude of the impact.
- Risk of accidental pollution.
- Environmental risks (the effect on human health and on aquatic flora and fauna).
- The continuity and level of pollution sources.

4. Description of the Integrated Water Monitoring System

As with any other activity, decisions in the field of water management cannot be made if the necessary information is unavailable. How good decisions are, and how quickly they can be made depends on the amount and quality of information available. The information needed for the decision-making process includes both present and past dimensions. Such information is useful for studying water management processes, which are stochastic processes, due to the very nature of meteorological and hydrological phenomena. Strategic and tactical decisions are based on such studies.

From the informational point of view, a distinction must be made between:

- Activity related to acquiring knowledge of natural phenomena; this activity generally covers the fields of meteorology and hydrology and is of a descriptive nature.
- Activity related to water resources planning, construction and operation of hydraulic structures, as well as to policies in this domain. This activity covers the field of water management and is of a decisional nature.

There is a close relationship between the disciplines of water management, meteorology and hydrology. Thus, even though each discipline has its own specificity, it not only provides information for its own purposes, but also for those of the other disciplines.

The transmission and concentration of data necessary to the decision-maker in the field of water management is realized by the informational system of water, which is based on the water dispatching offices (Figure 3). The offices are organised at the following levels and are interconnected by computer network:

- Local, for the most important local objectives.
- Regional (county) water management system.
- Basin, territorial branches of the Romanian Water Authority
- National, the Romanian Water Authority and the Ministry of Waters, Forests and Environmental Protection (MWFEP).

The information collected at the dispatching centres comes from the national meteorological, hydrological and water management networks and from international data networks according to the bilateral agreements signed with neighbouring countries and the Danube and the Black Sea Conventions. The transmission and concentration of hydrological data are done by the informational system in the field of hydrology, and comprise the following components:

- Rivers.
- Lakes.
- Groundwater.
- Sea.
- Precipitation.
- Evapotranspiration.
- Water uses.

In rapid flux 710 hydrometric stations and 1,700 rain gauge stations collect and transmit data, on the basis of which warnings and hydrological forecasts are worked out. In slow flux data are obtained from 1,016 hydrometric stations, 30 small basins (experimental and representative), 25 lakes for the study of silting, 904 complementary sections, 541 springs, 80 evapometric stations, 337 water uses stations and 4,582 hydrogeological wells.

On the basis of data obtained in the slow flux the annual hydrological and hydrogeological reports are made, and the amounts of water that were effectively drawn from the various sources are determined.

The informational system in the field of water management is made up of the following subsystems:

- Water quality: rivers, natural and artificial lakes, coastline, groundwater.
- Water uses.
- Accidental pollution.

- Operation of the hydraulic structures.
- Prevention against floods.

In rapid flux data are transmitted from 111 reservoirs and 65 water quality control sections and reported on a daily basis, and reports come in from a further 22 on a weekly basis. On the basis of these and other data provided by informational, meteorological and hydrological systems, reports regarding the state of the water management systems are made.

Operational decisions are generally taken on the basis of reports regarding the state of the water management system, worked out at the level of each Water Management System (county level) and then integrated at the basin level. A synthesis report is forwarded to the Romanian Water Authority and to the MWFEP. Under normal circumstances the report is made on a daily basis, but several times a day under emergency circumstances (floods, drought, accidental pollution, accidents at the hydraulic structures etc.).

The report regarding the state of the water management systems includes a written and a graphical part. The written part refers to:

- Meteorological diagnosis and forecast.
- Hydrological diagnosis and forecast.
- State of the hydraulic structures.
- Special situations in water supply.
- State of the water quality on the important rivers.
- Accidental pollution.

The graphical part includes:

- Water supply sketch for the important uses.
- Reservoir release rules .

In slow flux data are transmitted from 474 river water quality control stations, 210 reservoirs, 1,268 wells, 13 profiles on the Black Sea shore and 2,100 pollution sources. The water management annual report, the water balance, the water cadaster and water quality synthesis reports are based on data transmitted in slow flux. Both rapid and slow flux data are used as the background for the application of the economic mechanism in the water field.

5. Transboundary Water Monitoring

In order to know the state and evolution in time and space of the aquatic environment, and to be able to protect it, data and information are necessary on very large areas that may even exceed the boundaries of any one country. Transboundary water monitoring is realized on the basis of bilateral conventions with neighbouring countries and international conventions.

The objectives of transboundary monitoring are:

- To learn about the qualitative and quantitative parameters of the water resources that enter or exit one country's territory.

- To enable adequate protection measures for water quality and the aquatic environment to be taken.

- To be able to warn the neighbouring countries on the occurrence of extreme events such as flood, accidental pollution, or an accident at the hydraulic structures.

- Elaboration of the hydrological forecasts on the rivers that form or cross the boundary.

- To learn about the impact of hydraulic structures on the hydrological regime of the rivers at the border sectors.

- To enable any negative effects of water to be prevented or minimised.

- To ensure fair joint exploitation of some hydraulic structures.

The principles of transboundary data exchange are the following:

- Data referring to any type of information for the protection of people and goods from any negative effects of water are to be exchanged for free, according to the 40th Resolution of the World Meteorological Organization (WMO). Data exchange between countries for any other purposes should be done on an efficient and economic basis.

- Data exchange is usually done for the stations on the rivers that form the border of each state and/or for the stations near the border (corresponding stations) for transboundary rivers.

- Data exchange is done under normal circumstances to determine the amount of water resources that enter and exit one country's territory. In this case data from the corresponding stations are analysed jointly following certain procedures. Once agreed upon the data then become official. This data exchange is usually done yearly

- Under special circumstances such as flood, accidental pollution, or an accident at the hydraulic structures data are exchanged whenever needed.

- Data exchanged between countries should preferably have been obtained from the national monitoring networks and therefore should not require expensive supplementary processing either in terms of contents or form.

The exchanged messages and data refer to:

- Hydrological warnings regarding the occurrence of floods and accidental pollution

- Hydrological forecasts.

- Daily discharges and levels.

- Water quality parameters, usually one value/month. The parameters exchanged are usually temperature, pH, dissolved oxygen, CCO-Mng, BOD5, TDS, ammonium, nitrates, orto-phosphats, Fe, Mn, phenols, oil products;

- Data and parameters referring to the most important structures for flood control concern dams, lakes and deviations.

- Characteristic data of the water uses having a major impact on water resources.

Data should be transmitted according to the international codes and telecommunications procedures adopted by the WMO. Under normal circumstances:

- The levels and discharges are transmitted by a special code used by the WMO for the transmission of hydrological data, known as HYDRA code.

- The hydrological forecasts are transmitted by a special code used by the WMO, known as HYFOR code.

This information is transmitted daily through the computer network with telex lines as an emergency backup. Under exceptional circumstances, when the water level exceeds the limit, the data are measured every 1-3 hours (depending on the basin size) and transmitted free of charge, usually at 0100, 0600, 1200 and 1800 hours by Global Telecommunications System (GTS) system. If further information is required, then the country requiring the data will connect via computer to the database of the country where the information originates, and can obtain the data at a supplementary cost. In case of accidental pollution data transmission is done by computer network.

Romania participates in data exchange with neighbouring countries on the following basis:

- Convention on the Co-operation for the Protection and Sustainable Use of the Danube River, signed in Sofia on 29th, 1994.

- Convention on the Protection of the Black Sea against Pollution, signed in Ankara on 23rd March 1991.

- Bilateral Hydraulic Conventions with Hungary, Yugoslavia, Republic of Moldova and Ukraine.

Under the Danube Convention the following quality parameters are reported annually: temperature, pH, suspended load, dissolved oxygen, COO-Mn, CCO-Cr, BOD5, conductivity, TDS, Cl, SO4, HCO3, Ca, Mg, Na, K, NH4, NO2, PO4, Ptot, oil, Lindan, DDT, Fe, Mn, Zn, Cu, Cr, Pb, Ni, As, Cd, Phenols, Hg from 11 sections on the Danube of which 6 are situated in Romania and 5 in the upstream area of the Danube.Under the frame of the Danube Convention, the Danube Accident Emergency Warning System

(DAEWS) was established. It has a Principal International Alarm Centre (PIAC) in every country on the Danube.

The more important objectives of DAEWS are as follows:

- Protection of the population by protecting drinking water resources and water users in case of accidental pollution on the Danube or its tributaries

- Protection of the environment in general and of the water resources in particular against the effects of accidental pollution.

6. Conclusion

Integrated water monitoring is of strategic importance to Romania, and already a priority of the Romanian Government. Romania's transition to a free-market economy and the need for harmonisation with EU legislation are favourable factors for developing the present infrastructure and improving the economic bases in the water sector, all with the aim of improving the quality of life.

7. References

1. Groot S., Villass M., "Monitoring water quality in the future." Delft Hydraulic, 1995. "Proceedings of the International Conference on Aspects of Conflicts in Reservoir Development & Management", City University, London, 3-5 September, 1996.

8. Annexes

8.1 CRITERIA FOR EVALUATION OF POLLUTION IMPACT ON WATERS

Para meter	Characteristic/ criteria	Score for the classification			
		1	5	10	30
A	Dilution (Q_r/Q_c)	Moderate D>50:1	Low 10:1<D<50:1	Unacceptable 5:1<D<10:1	Accent. Risk D<5:1
B*	**Effluent quality** BOD5 (mg/l) CCO (mg/l) NH4(mg/l) Phenols (mg/l) Oil pr. (mg/l) Toxic subst. (heavy met., pest.etc.)	Moderate <60 <100 <20 <0.1 <0.4 X 1.25 CA	Low >60<160 >100<750 >20<100 >0.1<0.5 >0.4<3.0 X 2 CA	Unacceptable >160 >750 >100 >0.5 >3.0 X 5 CA	Accent. Risk >300 >1,500 >200 >1.0 >6.0 >X 10 CA
C	**Impact magnitude**	No impact	Moderate impact	Severe impact	Extreme impact
D	**Accidental pollution risk**	Reduced	Moderate	Critical	Extreme
E	**Environment risk** E1. Natural and biological diversity E1.1. Eutrophization E1.2. DO decrease E1.3. Sediments contamination E2. Health risk E2.1. Drinking water contamination E2.2. Others E3. Agriculture E4. Toxic persist.	Local risk Less urgent	Regional risk Less urgent Local risk Urgent	International risk Less urgent Regional risk Urgent	International risk Urgent
F	**Continuity and level of the pollution sources production**	Unknown Present or low	Probable Present or slightly increased	Confirmed Normal or increased	
TOTAL SCORE = A+B+C+D+E+F points **MAXIMUM SCORE** : 360 points **MINIMUM SCORE** : 12 points					

Notes: CA - effluent allowed concentration by permit
        Qr - river discharge
        Qc - effluent discharge (return flow)
        the most representative value is considered (usually toxic subst)

8.2 CONTACTS

The Ministry of Waters, Forests and Environmental Protection Responsible for Monitoring Water Quality:

- Contact: Mrs.Liliona Mara, General Director, Blvd. la Govt. of Nr. 12 Bucharest.
- Agency(ies) establishing and enforcing standards: ICIM Institute for Environmental Engineering Research Contact: Dr. Ion Ielev Director ICIM Street Spl Independenti Nr 292 Bucharest
- Scientific reviews are provided by scientists and other specialists from ICIM.

Data and Information on Tranboundary Monitoring:

- Romania's border countries are Hungary, Yugoslavia, the Ukraine, Moldova, Bulgaria
- 98.2% of Romania's land area is in the Danube basin.
- Its shared rivers are the Tisa, Somes, Crisuri and Mures with Hungary, the Danube with Bulgaria; the Danube tributaries Timis, Caras and Nera with Yugoslavia; the Siret and Prut with the Ukraine, and the Prut with Moldova.
- Romania has monitoring agreements with its neighbouring countries and monitoring sites as follows: 29 with Hungary; 18 with Yugoslavia; 10 with Moldova, and 17 with the Ukraine.
- The following parameters are measured monthly: level, discharge, precipitation, temperature, pH, CCoMn CBD5, dissolved oxygen, fixed residue, conductivity, chlorides, sulfates, calcium magnesium, sodium, potassium, nitrates, nitrite, orthophosphates, iron, magnesium, chromium, copper, biological, lead, zinc, phenol, detergents and total coliform.
- It is recommended that monitoring and assessment should be performed on the basis of ecological rather than political borders

151

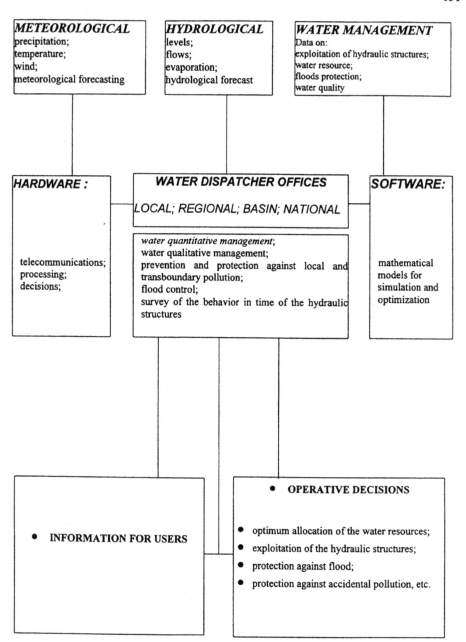

Figure 4. Schematic diagram of the informational - decisional system for water management

SLOVENIA

J. GRBOVIC
Ministry of the Environment and Spatial Planning of R Slovenia
Hydrometeorological Institute of R Slovenia
Ljubljana, Slovenia

M. BRILLY
University of Ljubljana
Faculty of Civil Engineering and Survey
Hydraulics Department
Ljubljana, Slovenia

1. Introduction

Slovenia is situated in Central Europe on the Adriatic coast. It has an area of 20,254 km², bordering Italy (232 km), Austria (330 km), Hungary (102 km) and Croatia (670 km). Its coastline on the Adriatic Sea is 46 km long. It has 2 million inhabitants, 90% of whom are Slovenes, and 10% members of Italian and Hungarian minorities and immigrants from other countries. Slovenia has three different climates: Middle-European, Alpine and Mediterranean. Slovenia varies in geological and tectonical aspects; 69% of its area are mountains and hills, and 43% of its area is influenced by erosion, because of ground structure, intensity of rainfall and temperature changes. The consequences of erosion are sediment transport and turbidity of water. 43% of Slovenia consists of karst, forests account for 53% of the country's surface, natural grasslands cover 24.7%, and arable land 12.1%. The main crops are wheat, maize and potatoes. There is also some viticulture in the river valleys.

Annual average intensity of rainfall varies from 750 mm/year in the northeastern plain areas of Prekmurje to 3,300 mm/year in the north in the Triglav National Park. The average yearly intensity of rainfall is 1,500 mm and average yearly runoff is about 1000 mm. 16500 km² of territory drains into the River Danube, and 3750 km² into the Adriatic Sea. The Slovene share of the Danube river basin covers about 81% of the country, and hosts about 80% of the total Slovene population. Slovenia receives 6% of the Danube River runoff from only 2% of the total Danube River watershed. Slovenia is rich in water resources, mainly groundwater and springs. Water from Slovenian territory collects in six main border and transborder water bodies: the Mura, Drava, Sava, Kolpa and Sofia rivers, and along the coast of the Slovenian part of the Adriatic Sea (Fig.1).

J. Ganoulis et al. (eds.), Transboundary Water Resources in the Balkans, 153–159.

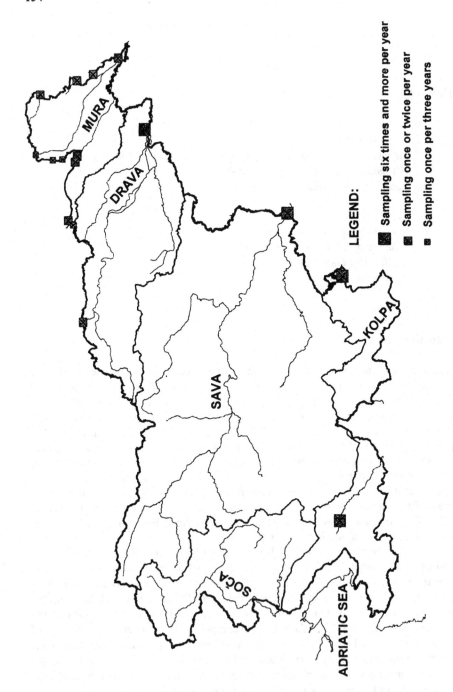

Figure 1 Main transboundary watersheds in Slovenia and transboundary sampling stations.

In the majority of cases, wastewater releases have caused a severe deterioration of water quality. It improved slightly in 1991-1998 as a result of reduced industrial activities. Domestic wastewater discharges are also important contributors. Unlike the surface water quality, the quality of the groundwater sources deteriorated in the first half of the 1990s. The nitrates and pesticides content in groundwater was especially high. In 1999 several municipalities were planning to upgrade their sewage treatment plants and 48.4% of inhabitants are connected to the sewage system (Environmental report 1995, 1996, 1998).

2. Environmental Legislation, Enforcement and Compliance (Slovenia, EU)

- The Environmental Protection Act (EPA) comprises the basic provisions regulating the protection of human existence and the inseparably linked natural environment as a constituent part of the regulation of development in the Republic of Slovenia. The EPA is based on Agenda 21, environment-related EU Directives, and Slovene experience with environmental management. The EPA forms the basis for the preparation of other legal instruments.

  According to the EPA, the monitoring of environmental pollution should include:

- The observation and supervision of immissions in the soil, water and air, the animal and plant worlds, and the health of ecological conditions (immission monitoring).
- The observation and supervision of emissions into the soil, water and air (emission monitoring).

Environmental monitoring is provided by the Ministry of the Environment and Spatial Planning in co-operation with the other Ministries (the Ministry responsible for agriculture and forestry, the Ministry responsible for health, the Ministry responsible for civil defence and rescue and the Ministry responsible for the natural heritage). The Minister proscribes in detail different issues pertaining to water monitoring (such as methodology of monitoring, its scheme, qualification of executors etc.). The Nature Protection Administration within the MoESP is responsible for qualitative and quantitative water management. The Hydrometeorologic Institute (HMIS) within the MoESP is responsible for the execution of water quality and water quantity monitoring.

In June 1996, Slovenia signed a European Agreement with the European Union (EU) and applied for full EU membership. This was also the official beginning in Slovenia of a long procedure of adjustment and approximation to legislative, economic and other aspects of the EU. The agreement contains provisions that will strongly affect environmental protection in Slovenia.

The EPA lays down a legal basis for fees and charges. The Government levies emission charges on wastewater and water use charges. Besides wastewater charges on water use the EPA includes provisions for concessions. These concessions are granted for different types of water use, such as fish farms, irrigation systems, small hydro plants and drinking water exploitation. The EPA established the Environmental

Protection Fund in 1990, which was essentially supported by wastewater discharge taxes, and electricity and gas charges.

The EPA requires environmental monitoring and the operation and maintenance of an environmental information system. Monitoring of natural phenomena and immission is the duty of the MoESP. The National Environmental Report is prepared by the MoESP and adopted by the National Assembly every year. The polluters themselves must accomplish emission monitoring.

3. Water Quality Monitoring in Slovenia

The monitoring of natural phenomena, pollution, and emissions into the environment (environmental monitoring) is carried out within the Republic of Slovenia. Monitoring of environmental pollution must include the observation and supervision of immissions in the soil, water, air, the animal and plant worlds, and the health of ecological conditions (immission monitoring). Monitoring of environmental pollution must also include the observation and supervision of emissions into the soil, water and air (emmission monitoring).

Slovenia has long had traditional co-ordinated monitoring programmes focused on the assessment of the environmental state of inland waters. Data have been available since 1965. Since 1982 all water quality data have been stored at the Computer Centre of the HMIS in integrated databases.

The immision monitoring of rivers in Slovenia has been performed since 1965 when the sampling sites network consisted of 20 locations, while today there are 120. Such monitoring was established according to the methodology recommended by the World Health Organisation (WHO) and the World Meteorological Organisation (WMO). The most important criterion applied in determining the sampling frequency and the necessary extent of analyses was the possible influence of surface water on groundwater and major springs, both used for water supply. Sampling is done in different seasons of the year, preferably at low to mean low discharges. Samples for all types of analyses at one location are taken simultaneously. Analytical quality is assured both by national intercalibrations and in tests within the EU Environmental PHARE Programme.

Rivers are classified into four quality classes according to the use of the water. The assessment of the category is based on an overall assessment of the different classification system used for chemical, microbiological and biological analyses. The saprobic index is grouped into 4 main and 3 intermediate water quality classes, including unpolluted or nearly unpolluted water up to badly polluted water.

The biocoenoses present in a stream reflect the overall ecological conditions at the sampling site in the stream, the quality of the water, as well as the physical environment (banks, riverbed, substrate, discharge, velocity, channel training, maintenance etc.). Apart from organic pollution the biocoenoses also reflect a variety of other stresses e.g. from inorganic pollution, toxic pollution, acidic pollution and physical changes (channel training, drought etc.).

Physical and chemical samples are collected at each sampling site in different seasons, two, six or twenty-four times a year. The parameters considered the most important pollution indicators are oxygen content, chemical oxygen demand (COD), biochemical oxygen demand (BOD), phenols, nitrogen compound, phosphorous compound, detergents, pesticides, mineral oils and trace metals. Field equipment is used to measure DO, pH, oxygen regime, conductivity and temperature.

Water samples are taken at the surface (0-0.5 m), preserved at the sampling site and transported to the laboratory for analysis by the next day at the latest. Grab samples (**sediments**) are taken close to the riverbank or from a boat. Samples for **biological** material are collected at the bottom close to the riverbank, from stones and from any solid surfaces in the water. Samples of periphyton may be scraped from different substrates in the water and macroinvertebrate fauna may be taken with the aid of a handnet. The biological material is sampled in the littoral of the flowing water down to a depth of ca. 0.5 m, where the sampling is not hampered either by water depth or speed. Biological samples are taken biannually, in the cold and warm season of the year, preferably at low to mean low discharges.

4. Monitoring of Transboundary Rivers

The Republic of Slovenia has several bilateral agreements with other countries for monitoring of immision on border and transborder rivers and creeks. The first of these covers monitoring on the Mura and Drava Rivers. It started in 1954-1955 under an agreement between Austria and the former Yugoslavia. The results of transboundary monitoring along the Mura River from 1989 to 1998 showed that the water quality improved to quality class II-III. This improvement can be attributed to the rehabilitation measures taken in Austria [8]. The situation is similar in the Drava River [9,10], where decreasing industrial pollution during recent years is considered to be the main reason for the decline in pollution. The Drava River is sampled once a year, the Mura twice a year and creeks are sampled once every three years (Figure 1).

Sampling and evaluation of data is done jointly by experts from both sides. The data include:

- **Physical and Chemical Parameters:** water+air temperature, pH, conductivity, DO, DO-sat. DO-def.,suspended solids, BOD5, COD, DOC, AOX, nitrogen compounds, phosphorous compounds, alkalinity, sulphate, chloride, detergents, phenol index, total hardness, carbonate hardness, noncarbonate hardness
- **Biological Parameters:** periphyton, macrozoobenthos (macroinvertebrate), total coliforms, faecal coliforms, faecal streptococci.

Transboundary monitoring with Italy includes only one underground karst, the Reka River that is sinking into Škocijanke Jame (Figure 1). The monitoring is under a 1984 bilateral agreement between the former Yugoslavia and Italy. Samples are collected six times a year and collected and analysed separately by experts from each country.

Transboundary monitoring of water quality between Slovenia and Hungary began in 1996 and includes the Velika Krka, the Ledava and the Kobiljski potok (Figure 1). Water quality data is collected jointly by experts from transboundary Slovenian-Hungarian monitoring teams.

Transboundary monitoring along the Croatian border used to be available and exchanged on national and international levels and this was adequate. However, in latter years this has not been the case. There are only two sampling sites, the Drava-Ormož and the Sava-Jesenice na Dolenjskem, both of which started in 1995 under the PHARE funded Environmental Programme for the Danube River Basin (EPDRB), when data were sent to a central database. In 1996 all EPDRB countries began operating the TransNational Monitoring Network and to further develop their monitoring and assessment programmes. Previously water quality data had been collected by experts from Slovenia and also separately by experts from Croatia, but as of 1999 experts from both countries have been working together on data collection. The water quality of the River Sava in Slovenia has also improved over the past years, but remains in quality class II-III

5. History of Transboundary Surface Water Quality Monitoring

Carefully drawn up procedures are essential for the management of water quality monitoring, especially for transboundary monitoring where two governmental institutions from different countries, with different practices, equipment and standards, are involved. Good practice should include:

- Common sampling of previously specified discharge, using specified common procedures
- Ensuring good and comparable results at sampling sites
- The common interpretation of results achieved under the same circumstances
- Joint final report

Friendly relationships and common experience are also very important for good final results, which should be scientifically presented to the public shortly after the measurements have been taken.

Comparability among the standards set for physical and chemical variables is relatively simple as (physico-chemistry) monitoring regimes are largely standardised. Methodology and procedures for analysing physical and chemical parameters are mainly standardised. Several methods may exist for particular parameters, but the experts from both sides have only to agree on which they will use for joint investigation purposes.

On the European level there is no commonly accepted, unique method for the assessment of biological flowing waters quality; certain procedures are standardised (e.g. biological sampling, presentation of biological quality data, [5,6,7]). Belgium and Italy, for example, have their own method of qualifying the quality of flowing waters, the results of which are only taken into consideration in Belgium and in Italy respectively. It is not currently possible to develop an EU-wide biological flowing water quality monitoring and assessment system, but such a system would solve many of the problems of comparability. The European Commission is determined to ensure that the research necessary for this takes place in the medium term (Council Directive, establishing a framework for community action in the field of water policy COM (97) 49 final).

Slovenia's experience with transboundary monitoring between Slovenia and Austria shows that in biological flowing water quality monitoring it is important that the group of experts should remain the same for a few years. Working together has allowed them to train together and to unify procedures as well as criteria. This means that the results are more reliable, which is particularly important when monitoring takes place over a long period of time.

6. References

1. Draft Country Report for 2nd Meeting of the SECI Project Group from the Republic of Slovenia. Prepared by Ecological Engineering Institute, Maribor, Slovenia
2. Environmental Performance Reviews SLOVENIA. Economic Commission for Europe, Committee on Environmental Policy, United Nations Publication, Sales No. E.97II.E.16, N.Y. and Geneva, 1997
3. Pantle, R. & Buck, H. Die biologische der □berwachung der Gew□sser und die Darstellung der Ergebnisse, *GWF*, 1955, **96**, 604.
4. Zelinka, M., Marvan, P. Zur Pr□zisierung der biologischen klassifikation der reinheit fliessender Gew□sser *Arch. Hydrobiol.*, 1961, **57**, 389-400.
5. ISO 7828 (E) Water quality - Methods of biological sampling Guidance on handnet sampling of aquatic benthic macroinvertebrates, 1985.
6. ISO 8265 (E) Water quality - Design and use of quantitative samplers for benthic macroinvertebrates on stony substrata in shallow freshwater, 1988.
7. ISO/DIS 8689-2 Water quality - Biological classification of rivers-Part 2: Guidance on the presentation of biological quality data from surveys of benthic macroinvertebrates, 1998.
8. Standige Ostereisch-jugoslawische Kommision f□r die Drau. Gemeinsamer Bericht □ber die Untersuchung der Wasserg□te der Mur im Grenzbereich, MoESP, 1965-1991
9. Standige Ostereisch-slovenische Kommision f□r die Drau. Gemeinsamer Bericht □ber die Untersuchung der Wasserg□te der Mur im Grenzbereich, MoESP, 1992-1998
10. Brilly, M., Grbovic, J., Gorišek, M. Crossborder monitoring of water quality in Slovenia. I. Lyons Murphy (ed.), *Protecting Danube River Basin Resources*, 189-198. Kluwer Academic Publishers, 1997

TURKEY

A SAMSUNLU, A. TANIK & L. AKCA
ITU, Istanbul Technical University
Faculty of Civil Engineering
Maslak, Istanbul, Turkey

1. Introduction

Turkey is located in the southeastern part of the Balkans. It joins the two continents of Europe and Asia. The European part is named Thrasian and the Asian part Anatolia. The country is surrounded by sea on three sides; the Black Sea in the north, which is shared by six riparian countries, the Aegean Sea in the west and down through the Marmara Sea to the Mediterranean Sea in the south, as shown in Figure 1. The total shoreline is approximately 8,300 km long and half of Turkey's population of 65 million lives in coastal cities, districts and villages, even though the coastal area covers only 29 % of the total surface area of the country. The Thrasian part of Turkey is small but has a fairly high level of population density, as part of Greater Metropolitan Istanbul is located here.

Rapid urbanisation and industrial development is responsible for most of Turkey's environmental pollution. While only 18 % of the population lived in urban districts in the fifties, in recent years this figure has reached 67 %. As a result of industrialisation, the ratio between rural and urban populations has changed rapidly, especially since the 1980s. The population in highly touristic coastal areas has grown even faster than in other areas. Additionally, the population increases even more during the summer season. Due to rapid growth in urban and touristic areas and to industrial development, the way in which land is used has drastically changed. As a result, environmental pollution has occurred especially in coastal areas. Infrastructure is needed in order to prevent further pollution. Wastewater collection and disposal systems have been planned and constructed in some areas, but are still insufficient elsewhere. All municipalities have their own water supply systems. Wastewater disposal is more expensive than water supply. As Turkey is a developing country, it has a very limited budget for environmental protection. Any funds that are available should be spent efficiently.

J. Ganoulis et al. (eds.), Transboundary Water Resources in the Balkans, 161–174.
© 2000 *Kluwer Academic Publishers.*

162

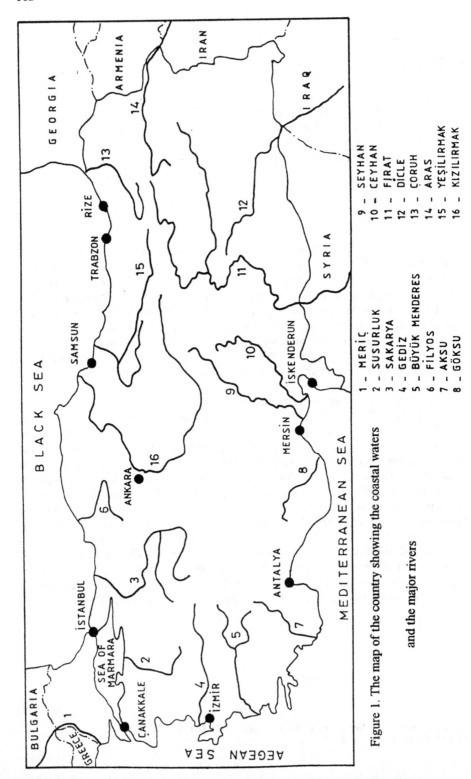

Figure 1. The map of the country showing the coastal waters

and the major rivers

1 – MERİÇ
2 – SUSURLUK
3 – SAKARYA
4 – GEDİZ
5 – BÜYÜK MENDERES
6 – FİLYOS
7 – AKSU
8 – GÖKSU

9 – SEYHAN
10 – CEYHAN
11 – FIRAT
12 – DİCLE
13 – ÇORUH
14 – ARAS
15 – YEŞİLIRMAK
16 – KIZILIRMAK

Almost 60 % of economic activities in Turkey take place in coastal areas, because of the favourable climatic conditions and their proximity to international waters. According to annual average values, the country receives a precipitation of 630 mm. At Rize, located on the northeast coast (see Figure 1), a maximum value of 2,000 mm and minimum of 300-350 mm have been recorded. In terms of economic strength, Turkey lies in 16th position in the world with an annual national product of 215 billion US $.

The most important transboundary river in the Thrasian part of Turkey is the Maritza River, which forms the border with Greece and is shared by Greece and Bulgaria. The other transboundary river is the River Tundja, which is shared by the same three countries. Although in Anatolia Turkey shares other transboundary rivers with Middle East countries like Iran, Iraq and Syria, these rivers are excluded from this paper as the framework of this study is limited to Balkan countries.

In general, Turkey's major water supply comes from groundwater, however, in the biggest three Greater Metropolitan cities of Istanbul, Ankara and Izmir, drinking water is supplied from surface waters. In such rapidly extending and growing regions ground water resources have become insufficient and therefore surface waters offer a solution to water demand. Water is supplied from rivers through constructed dams and reservoirs, but water is urgently needed, as there are signs of pollution. Industrial water needs are also supplied from rivers; however, reuse has not been practised so far. Despite recent industrialisation, Turkey is still basically an agricultural country. Approximately 40% of the population deal with agricultural activities, and irrigation water for arable land is supplied through irrigation channels receiving water mainly from rivers.

2. Water Monitoring

2.1 ENVIRONMENTAL ADMINISTRATIVE AND LEGAL STATUS

A brief outline of the legal status on environmental pollution control, protection and monitoring is given below:

- Law on Aquatic Products, 1973.
- Environmental Affairs, a responsibility of the Under-secretary of the Ministry of State Works, 1978.
- Environment Law, 1983.
- Turkish Water Pollution Control Regulation, TTWPCR, 1988.
- Establishment of the Ministry of the Environment, 1991.

Water quality and quantity measurements of surface waters like rivers are done by the State Water Works (DSI) with 16 regional administrations around the country, and by the Administration of Electricity Surveys (EIEI) under the Ministry of Energy of the Republic of Turkey, respectively. The monitoring work of the coastal waters is mainly conducted by the Ministry of the Environment (water quality) and by the Ministry of

Agriculture (fishery). Turkey has a central autonomy. There are 80 provinces in the country and the administrative works of these provinces are governed from Ankara through related authorities under the Ministries. Related ministries have Provincial Directories in each of the provinces under the Governor of the provinces. Some of the major rivers are the responsibility of regional directors as they are monitored on a catchment area basis under the responsibility of General Directorates of related ministries.

The regional authorities are the municipalities. *Metropolitan Municipalities* gained power under the Municipalities' Law number 3030. These municipalities now have more freedom to solve their own environmental problems. They are able to collect money from residents for both the drinking water supply and for wastewater treatment and disposal. Thus, they can construct sewerage collection systems and marine discharge systems, and some of them have even already managed to install and operate wastewater treatment plants. Therefore, money paid by the inhabitants has started to show a return in the form of service. They have also urged industries to construct their own industrial treatment plants. There are almost 16 metropolitan municipalities in the country, some of which are situated along the coasts e.g. Istanbul, Izmir and Antalya.

Other municipalities still come under Municipalities Law number 1585 (created in 1933) and are economically quite weak. There are about 2,850 such municipalities and they do not have the freedom to collect money from residents. The central government organises all activities aimed at solving these municipalities' environmental problems, including finding money even from foreign associations. Thus the government determines the priorities regarding environmental pollution control. Higher priority areas are those situated along the coasts; thus the construction of sewerage and marine discharge systems in these areas has been accelerated, as mentioned in Samsunlu et al[1].

2.2 CURRENT WATER QUALITY STANDARDS

The regulation on the quality of surface waters like rivers, dams and lakes is based on the Water Quality Classification of Inland Surface Waters entitled Water Pollution Control Regulation (TWPCR) [2]. Parameters for the classification are grouped under four headings:

- Physical and inorganic chemical parameters.
- Organic parameters.
- Inorganic pollution parameters.
- Bacteriological parameters.

The regulation is based on the 90 percentiles of pollutant concentration data in water. The lowest quality value determines the quality class for a given group of parameters. According to TWPCR, water bodies evaluated as such are grouped in four classes and their proposed beneficial uses may be summarised as follows:

2.2.1 First Class – High quality water
Following disinfection water can be used for drinking water supply, recreational purposes including body contact sports like swimming, trout production, animal breeding and farm use, and other purposes.

2.2.2 Second Class – Slightly polluted water
After pertinent treatment water can be used for recreational purposes, for fish production other than trout, and for irrigation, provided that the standards given in the Technical Methods Circular are complied with, and for other uses except those outlined for Class 1 water.

2.2.3 Third Class – Polluted water
For industrial use except for the food and textile industries, which need high quality water.

2.2.4 Fourth Class – Highly polluted water
Other uses allowing low quality water.

Within this context only water from Classes 1 and 2 may be used for supply purposes after prescribed treatment.

The regulation on coastal waters and marine waters are given separately in the form of Tables in TWPCR. The required standards on coastal waters used for recreational purposes are given in Table 3 of TWPCR [2]. The parameters used are colour, taste and odour, light intensity, pH, oil and grease, total coliform, faecal coliform, surfactants, phenols, dissolved oxygen, floatable matter and tar residues. The general quality criteria of marine environments are stated in Table 4 of TWPCR [2]. The parameters taken into account are colour and turbidity, floatable materials, suspended solids, dissolved oxygen, degradable organic pollutants, raw petroleum and petroleum derivatives, radioactivity, productivity, toxicity, phenols, various heavy metals (copper, cadmium, chromium, lead, nickel, tin, mercury and arsenic) and ammonia.

The authorities in charge of water resources, both in terms of quality and quantity, and of the design and construction of water and wastewater treatment facilities are the Bank of Provinces (under the Ministry of Public Affairs), the Ministry of Forestry, the Ministry of Tourism, the Ministry of the Environment, the State Universities and the Turkish National Scientific and Technical Research Centre (TUBITAK).

2.3 MONITORING STUDIES CONDUCTED AT COASTAL WATERS

2.3.1. The Black Sea

- Studies on determination of fish carrying capacity, annual stock data, (TUBITAK, Ministry of Agriculture).

- Periodical measurements around marine discharge points at pre-determined stations (Bank of Provinces).

- Pollution Status of the Black Sea, Global Environmental Facility (GEF) Black Sea Environmental Programme (Ministry of Environment, Ministry of Health, Ministry of Foreign Affairs).
- Pollution determination around the Yesilirmak River

2.3.2 Sea of Marmara

- Determination of fish carrying capacity, annual stock data (TUBITAK, Ministry of Agriculture).
- Search for marine outfall possibilities and applications around Istanbul (ISKI [8], General Directorate Water and Sewerage Works of Greater Istanbul Municipality, Regional State Universities).
- Pollution profile surveys of the sea (Regional State Universities).

2.3.3. Aegean Sea

- Pollution measurements all around the sea (Bank of Provinces).
- Determination of fish carrying capacity, annual stock data (TUBITAK, Ministry of Agriculture).
- Search for marine outfall possibilities and applications around the coastal area (State Universities).
- Various national and international projects on coastal zone management.
- Pollution profile surveys of the sea (Regional State Universities, Institute of Marine Sciences, Izmir).

2.3.4. Mediterranean Sea

- Determination of fish carrying capacity, annual stock data (TUBITAK, Ministry of Agriculture).
- Search for marine outfall possibilities and applications around the coastal area (Bank of Provinces, State Universities).
- Pollution profile surveys of the sea (Regional State Universities, Middle East Technical University (METU), Institute of Marine Sciences, Mersin).

In this study, emphasis will be given to Turkey's coastal waters in terms of transboundary Balkan water resources, as they constitute the major part of the subject matter. General information on the coastal waters, their current pollution levels, and wastewater collection treatment and disposal systems on Turkey's coasts will be briefly outlined. As the nature of the pollution problems differ from region to region, the problems will be discussed on the basis of their nature.

3.   The Current Status of Transboundary Coastal Areas

The seas surrounding Turkey are generally distinct from each other (Figures 1 and 2, satellite view). This restricts water exchange, and it is thus more difficult to dilute or flush wastes discharged to these seas. Also, vertical mixing of water masses stop below a certain depth (especially in the Sea of Marmara and the Black Sea), which concentrates pollutants within each stratum. The cause of the separation varies. The narrow straits of Istanbul and Canakkale that connect the Black Sea to the Sea of Marmara, and the Sea of Marmara to the Aegean Sea respectively, block the confluence of these waters. The Aegean Sea's connection to the rest of the Mediterranean is constrained because of many islands, including Crete and Rhodes.

The Aegean and the Mediterranean coasts of the country are almost the cleanest parts of the Mediterranean Sea. These regions (south and west coasts) are Turkey's most important touristic areas. Thus, these coasts should be considered as environmentally sensitive zones. There are considerable amounts of agricultural and industrial production as well as tourism in these areas. On the other hand, population density is also higher compared to other regions. Sewage, industrial discharges, and diffuse load from agricultural areas are conveyed to the sea environment through rivers and streams.

Each coastal zone has different features. Information on each sea is briefly presented below [3].

3.1 THE BLACK SEA

The Black Sea extends over 1,200 km from east to west and 600 km from north to south. With a base of 422,189 km^2, it is a semi-inland sea. Total water volume is 536,969 km^3, 87% of which is deep water without oxygen.

- The water collection basin is 2.2 million km^2 and the average depth is 1,272 m; 37 % of the sea floor is deeper than 2,000 m. The shallow sections (less than 200 m) are to the north and west and constitute 27 %.

- Average salinity is 0.18-0.19 %, increasing below 100 m and reaching 0.22 % around 200 m.

- The sea has a surplus of surface waters because of high precipitation, limited evaporation and the abundance of continental fresh water inflows. This leads to an annual average outflow of 612 km^3 of surface water into the Sea of Marmara. However, the sea also receives a 312 km^3 average annual inflow of saline water from the Mediterranean through the counter current along the Strait of Istanbul.

- The pollution load is relatively high from natural causes and waste deposited from large rivers from several countries (including Turkey). Pollutants flow in from 16 countries and 160 million people live in its catchment basin.

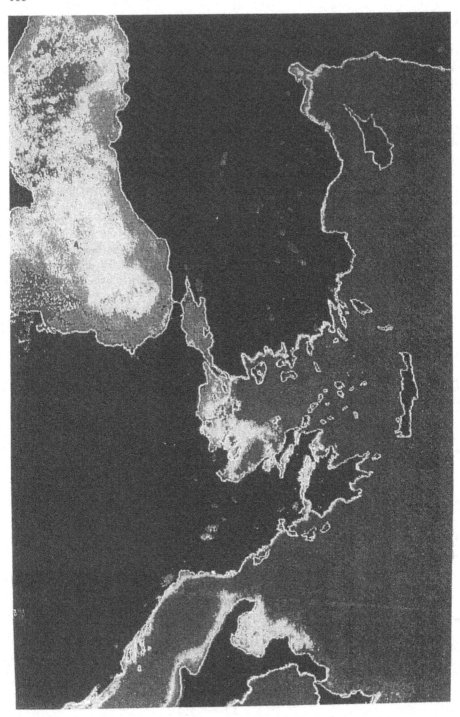

Figure 2. Satellite view of the seas surrounding Turkey

- The Danube River alone discharges 60 tons of mercury, 1,000 tons of chromium, 4,500 tons of lead, and 50,000 tons of oil annually, which in turn affect the Sea of Marmara and the Aegean Sea.

It is also an aquatic environment where bacteria found in sewage can remain alive longer than in Turkey's other seas due to relatively low solar radiation, water temperature and salinity. The sea is rich in plankton and fish that live on this biomass making it Turkey's most important fishery. However, catches have been declining due to overfishing and the sea's changing ecosystem, resulting from a new ctenophore (Mnemiopsis) introduced through the ballast water of ships. The damage to fish hatcheries may be occurring due to shoreline erosion. Fishing potential has dropped from 580,000 tons a year in 1988 to 290,000 tons in 1991, while it recovered to around 550,000 tons in 1995[4].

The Black Sea receives large quantities of mostly untreated domestic and industrial wastewater. Turkish coasts of the Black Sea receive a considerable quantity of fresh water via streams and rivers. Some of the large rivers (Kizilirmak, Sakarya and Yesilirmak) carry industrial and municipal discharges. The variety of industries range from food manufacturing to cement factoring. The basic industrial categories are food manufacturing, the paper and paperboard industry, wood products, the fertiliser industry, the chemical industry, basic metal industries and the beverage industry. Furthermore, as expected, the Black Sea coasts have continued to develop rapidly. The Black Sea is further stressed by an increased input of waste from new residential and industrial areas. The topographic situation of the Black Sea region affects the waste discharge points. Since the mountains are parallel to and generally very near the coast, there is only a very narrow band of land where people can settle. So, especially in the east part of the Black Sea, there are dense population settlement areas from which sewage is discharged directly in a continuous band to the Black Sea.

The pollution loads from the Danube, Denieper and other streams and sources affect water quality on the Turkish coast of the Black Sea. The counter-clockwise circular current in the western part of the Black Sea plays an important role in the carrying of pollutants within the sea itself.

The Bank of Provinces (Iller Bank) has started to build wastewater treatment plants and sea outfalls at many places, starting with big settlement areas. Industrial plants usually build their own wastewater treatment plants, which are controlled by the Ministry of the Environment. Table 1 shows the current situation regarding wastewater treatment plants and sea discharges.

TABLE 1. Wastewater treatment plants and plants for discharges into the sea in the Black Sea Region [5].

	Biological		Sea Outfall	
	Number	Population	Number	Population
In service	2	45660	5	203654
Under construction	2	147503	11	554712

3.2 THE SEA OF MARMARA

The Sea of Marmara also displays peculiar hydrodynamic features because of the structural character of the straits that connect it to other seas.

- It extends over 11,350 km^2 and has 3,377 km^3 of water. Waters of Black Sea and Aegean Sea origin form two distinct strata. In the south there is a wide and relatively shallow shelf, but to the north depths are over 1,000 m extending from east to west.

- Its oceanographic features vary in line with those of the Aegean and Black Sea, as it is connected to these two seas. Water temperatures vary according to seasons. Less saline (0.16-0.18%) surface waters of Black Sea origin and the deep waters of Aegean origin with higher salinity (0.38-0.39%) form two strata with limited confluence.

- The oxygen saturation of waters at 25-30 m varies from 20-30 %, which poses a problem in the decomposition of coastal discharges and organic substances coming from the Black Sea.

- An estimated 766 million m^3/year of wastewater are discharged into the sea; this figure does not include wastewater from industries in metropolitan Istanbul, since no detailed inventory exists.

Organic matter equal to 158,000 tons of BOD and 370,000 tons of COD are discharged annually into the surface waters of the Strait of Istanbul at its junction with the Sea of Marmara. Eutrophication declines in phyto- and zooplankton. The disappearance of certain fish species and significant algal growth has been noted in recent years. One species of algae, Gracilaria, has grown enormously and is being collected commercially for export [4].

Critical coastal and near-shore areas include:

- The Bay of Izmit, which receives waste from Turkey's most important industrial area as well as the domestic waste of the city of Izmit.

- Gemlik Bay, which receives pollution from Lake Iznik as well as industrial and household waste from adjacent towns via Golayagı Stream.

Istanbul is the largest city located by the Marmara Sea. The other large and industrialised cities are Bursa and Izmit. Roughly 70% of Turkey's industry is located on the Marmara coast. Istanbul is also Turkey's most important commercial, touristic and cultural centre. The water quality and the pollution load to the Sea of Marmara from the surrounding area has been investigated within the Istanbul Master Plan study [6]. There is a two layer current system in the Bosphorous. One is the dense Mediterranean water flowing to the north at the bottom, the other is brackish surface water from the Black Sea carrying runoff from inflows, such as the considerable amount of organic carbon, nitrogen and phosphorus from the Black Sea to the Marmara Sea. Since 1998

63% of Istanbul's domestic wastewater is discharged after it is treated. The aim is that by the year 2000 95 % of domestic wastewater will be treated before it is discharged [7].

Due to the lower layer flow from the Marmara Sea to the Black Sea, all the discharges to the Bosphorus have been pretreated. The short-term plan is that all discharges to the Marmara Sea will have been previously biologically treated, and that in the long-term there will be nutrient removal. According to 1997 data, 82.6% of the wastewater of industrial plants in Istanbul is treated physically, chemically or biologically before being discharged into the sewer net system or receiving water [8].

In Izmit, which is the second important city by the Marmara Sea, there is one biological treatment plant already in use for municipal wastewater. Another is under construction. There is also a treatment plant being used for the treatment of industrial wastewater. More than 90% of the industrial plants in Izmit Bay discharge their previously treated wastewater into the sea or the sewer system. Table 2 shows the number of wastewater treatment plants and sea discharges.

TABLE 2. Wastewater treatment plants and plants for discharges into the sea in the Marmara Region [5]

	Biological		Sea Outfall	
	Number	Population	Number	Population
In service	2	357799	8	634796
Under construction	5	523300	2	42702

3.3 THE AEGEAN SEA

- With a coastal perimeter of 2,805 km, the Aegean Sea is one of the five distinct basins of the Mediterranean. Its length from north to south is 660 km, while its width is 270 km. in the north, 150 km. in the centre, and 400 km. in the south. Its surface area is 214,000 km^2 and its average depth is 100-150 m. Due to currents, its water movement is extremely varied.

- Wastewater discharges into the Aegean Sea occur at nearly 50 major points along the coast (seven rivers, at least 40 tourism and vacation home developments, one industrial zone, and input from the Black Sea) as well as a number of domestic sewage outfalls. The total pollution load from these sources is equal to a population of 20 million, 10 million of which comes from the Black Sea discharges. To this, a load of 7.5 million population equivalents (PE) should be added from adjacent Greek settlements and industries for a total load equal to that of 27.5 million people. Localised pollution problems include high levels of suspended solids, dissolved/dispersed petroleum hydrocarbons, mercury, and cadmium. BOD, nitrogen and phosphorus from sewage discharges in the northern Aegean are expected to nearly double from 1990-2010.

Critical coastal and near-shore areas include:

- The Bay of Izmir, whose inner bay suffers from organic pollution from Izmir's sewage, from heavy metal concentrations from industries, and petroleum and other ship wastes from port activities.

- The Candarlı Bay, which is polluted by tanker traffic, refineries and tanker-filling installations, as well as by organic loads from the Bakircay and Buyuk Menderes Rivers.

- The coastline from Kusadası to Marmaris, which has pollution problems and a disrupted ecosystem caused by rapid population and tourism growth, and the construction of secondary homes.

The Tourism, Infrastructure and Coastal Zone Management Project is one of the most important projects in this region and is supported by the World Bank under the sponsorship of the Ministry of Tourism. All types of issues related to the environment were considered, investigated and evaluated. These issues included tourism, agriculture, socio-economics, hydrogeology, water environment, nature, infrastructure, current legislation and financial conditions. Recommendations for solving the various problems were then made using a master plan approach. Aliaga Refinery and other industries have their own treatment plants. Table 3 shows the wastewater treatment plants and discharges into sea.

TABLE 3. Wastewater treatment plants and plants for discharges into the sea in the Aegean Region [5]

	Biological		Sea Outfall	
	Number	Population	Number	Population
In service	8	727084	13	301249
Under construction	6	605575	6	118737

3.4 THE MEDITERRANEAN SEA

With a surface area of 2.5 million km^2, the Mediterranean is the largest inner sea in the world. The northeastern Mediterranean has a narrow continental shelf; oxygen exists at all depths; a westerly current moves along the Turkish coast; average salinity is around 0.38 %; average annual water temperature is 15-17° C; and the total annual discharge to the sea is 36,300 m^3, entirely from streams. Because of coastal settlements and high maritime traffic, this part of the sea is very sensitive to pollution.

Annual discharge into the sea from rivers and sewage channels is 36.3 billion m^3, 99 % of which are river effluents. Although industrial wastewater constitutes less than 1 % of the total volume, this contains highly toxic substances such as mercury, lead, chromium, and zinc[4]. Agricultural activities constitute the largest volume of pollutants carried to the sea by rivers and streams: 90 % of tobacco and sunflower seed production, 80% of cotton and corn output and 70 % of rice growing occurs in the coastal provinces [9]. Farming activities contribute 58 % of COD, 29 % of phosphorus, 24 % of nitrogen, and 14 % of BOD.

Critical coastal and near-shore areas in Turkey along the Mediterranean include:

- The Bay of Iskenderun, which has both special hydrological features (shallow bottom, upwelling, potential for aquatic products) and an inflow of domestic and industrial wastewater.

- The coast from Kemer to Alanya (including the city of Antalya), which has recently experienced rapid population and tourism growth that, in turn, have overloaded the environmental infrastructure and disrupted the ecology (from construction along the coast).

- The Goksu delta, which is a specially protected area because of its value for waterfowl preservation and reproduction.

- Various shoreline areas where secondary homes and tourist facilities are densely developed.

The sea is also home to the Mediterranean monk seal, which is one of the 12 most endangered species in the world. Of the 300-400 thought to exist, about 50 live on desolate parts of this coast.

In the northeast Mediterranean, most of the land based pollution loads come from Turkey. The pollution load consists of agricultural runoff, domestic and industrial wastewater discharges and the pollution carried by streams. The petroleum refinery in Mersin is one of the most significant pollution sources. Rivers and streams are the most important source of sea pollution in the region. For example streams carry 75% of BOD_5.

On the Mediterranean coasts, places like hotels, motels and clubs treat their own wastewater before discharging or using for garden irrigation. In such places biological package treatment systems are usually used. In addition, the Bank of Provinces has planned and constructed wastewater treatment plants and sea discharges for many places. The wastewater treatment plants and sea discharges that are under construction or in use and the population that they serve are shown in Table 4. Mersin Atas Refinery, Iskenderun Iron-Steel Plants and most of the industries in Mersin and Adana have their own wastewater treatment plants.

TABLE 4. Wastewater treatment plants and sea discharges into the sea in the Mediterranean Region [5]

	Biological		Sea Outfall	
	Number	Population	Number	Population
In service	3	118976	2	112787
Under construction	1	58104	1	46295

3.5 OVERALL QUALITY OF THE TRANSBOUNDARY WATERS

Figure 2 shows the satellite view of the transboundary coasts surrounding Turkey. The distribution of plankton can be detected in this satellite image indicating the density of plankton by using different colours. Their concentration is the lowest in the Aegean and Mediterranean coasts, however they show high densities in the Marmara and Black Sea regions. The water in the Black Sea is heavily polluted due to inflow from the Danube River. The Sea of Marmara is also polluted due to pollution from the Black Sea in the flow of the upper region of the sea, and partially due to pollution load from Istanbul and Izmit.

The dark colour indicates that chlorophyll concentrations exceed 1µg/lt., whereas the lighter colour indicates that the concentration is below 1µg/lt., [6].

4. Conclusions

This report on Turkey concerns itself mainly with the transboundary waters surrounding Turkey on three sides. General information is given on the Black Sea, the Sea of Marmara, the Aegean Sea and the Mediterranean Sea. The current status of the Turkish coasts is summarised together with a general background information on the country in terms of water resources. The paper also makes reference to legal and administrative issues concerning the environment, to current water quality standards, and to recent monitoring studies conducted in Turkey, with special emphasis on the overall water quality of transboundary waters.

Rapid urbanisation and industrial development are responsible for most of Turkey's pollution. Where residential areas have been established in coastal areas, a significant level of pollution is also evident. During the last decade efforts have been made to improve infrastructure and to apply methods of coastal zone management.

5. References

1. Samsunlu, A., Tanik A., Maktav, D., Akca, L., and Uslu, O. (2000) Coastal Zone Management Applications In Turkey, NATO ASI Series
2. TWPCR. (1988) Turkish Water Pollution Control Regulation, Turkish Federal Register, September 4, 1988.
3. NEAP. (1998) National Environmental Action Plan, State Planning Organisation, Ankara
4. EFT. (1995) Environmental Profile of Turkey, Environmental Foundation of Turkey, Ankara.
5. Iller Bank (1998) The Bank of Province of Turkey, Special Information, Ankara.
6. IMC (1997) Istanbul Water Supply and Wastewater Disposal Master Plan, IMC (in Turkish).
7. Eroglu V., and Sarikaya, H.Z. (1998) Achievements Towards Better Supply and Wastewater Disposal in Istanbul, Proceedings of Int. Symp. On Water Supply and Treatment, pp 1-20, Istanbul.
8. ISKI (1998) Istanbul Water and Sewerage Administration, Annual Report, Istanbul Metropolitan Municipality.
9. OECD (1992) Turkey's Environmental Policies, Paris.

FR OF YUGOSLAVIA

D. LJUBISAVLJEVIC
Faculty of Civil Engineering
University of Belgrade, Yugoslavia

S. VUKCEVIC
Institute for Technical Research
University of Montenegro
Podgorica, Yugoslavia

1. Institutional Framework for Water Pollution Control in the Federal Republic (FR) of Yugoslavia (Serbia And Montenegro)

The legal framework for the protection of water resources and the environment in general in the FR of Yugoslavia establishes the responsibilities of government bodies, institutions and other organisations with respect to the implementation of legal, planning and programme-related activities. Rivers in FR of Yugoslavia are shown in Figure 1.

Federal and republic regulations that create the water pollution control legal framework stipulate the competencies, obligations and responsibilities of the relevant institutions. In accordance with legal powers and responsibilities, quality control of surface waters of inter-state and inter-republican watercourses, as well as watercourses intersected by the state border is performed by the Federal Hydrometeorologic Institute, while the control of other watercourses is performed by the relevant republican Hydrometeorologic Institute. The federal and republican governments each year adopt the programmes of systematic water quality control. The above mentioned Hydrometeorologic Institutes also monitor water levels and flows on the rivers, and maintain data bases on the quality of waters, water levels and flows.

Quality control of surface waters in the sectors of rivers and impounding reservoirs, used as a source of water supply or bathing place, is carried out by the health service, but not according to the above-mentioned programme, but according to the programmes of the municipalities and towns where the water source or bathing place is located. In addition to the control of raw water intended for domestic water supply, regional public health institutes also perform the control of drinking water quality. The results of control are submitted to the inspection authorities and to the Republican Public Health Institute. The health service keeps evidence of all diseases caused by poor quality of drinking water.

J. Ganoulis et al. (eds.), Transboundary Water Resources in the Balkans, 175–184.
© 2000 *Kluwer Academic Publishers.*

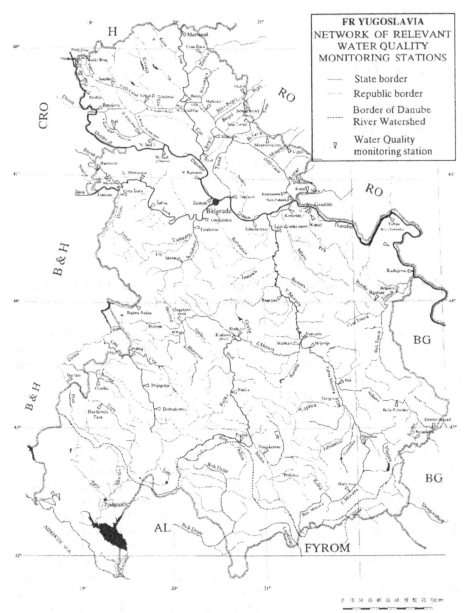

Figure.1 Rivers in Yugoslavia.

Control of the composition of wastewater, as well as the assessment of its impact on the recipient, is carried out by the enterprises and institutions authorised by the competent ministry, since they meet the requirements with respect to personnel, equipment and facilities. The results of such tests are submitted not only to the client, but also to the Republican Hydrometeorologic Institute, which then creates the appropriate database.

The quality standards for surface waters and waters intended for irrigation are jointly adopted by the ministries charged with water management, agricultural, environmental and health issues, while drinking water standards are adopted by the Federal Ministry of Health.

However, there is still not sufficient co-operation among the departments authorised for specified segments, because the terms of reference are not clearly defined. This means that the activities of, say, 3 or 4 ministries overlap, since each is responsible for one segment (the Ministry of Water Management for surface water quality, the Ministry of Mining for ground water quality, the Ministry of Health for drinking water and the Ministry for Municipal Services for the condition of plant and piping in water supply systems). The situation is similar with respect to chemical accidents caused by vessels, where the ministries of internal affairs, water management, health, environmental protection and transport are all involved.

2. Current Policies and Strategies

In all the above mentioned strategic documents it is stated that the development of individual economic sectors and the country as a whole should take environmental protection into consideration and at the same time improve the current state of affairs, especially in the area of water protection and water management in general.

If the FR of Yugoslavia joins the international financial markets, and its economy strengthens soon, it can be expected that the most of the strategic aims will be achieved, if not within the anticipated time limit, then certainly in the near future.

2.1 WATER QUANTITY AND QUALITY MONITORING

Water quantity and quality monitoring is done by the Hydrometeorological Institute of the Republic of Serbia and the Hydrometeorological Institute of the Republic of Montenegro. A summary of monitoring activities is given below:

In the Republic of Serbia physico-chemical, biological and microbiological analysis are done on:

- 156 profiles on 83 rivers once a month
- 23 profiles on 14 rivers once a week
- 10 profiles on 8 rivers once a day
- 34 resources once a year
- 33 reservoirs resources once a year
- 97 groundwater resources once a year

In the Republic of Montenegro physico-chemical, biological and microbiological analysis are done on:

- 20 profiles on 9 rivers 7 times a year.

The Federal Hydrometeorologic Institute co-ordinates the activities of the Republic Institutes and is responsible for international co-operation.

As well as flow rates, further physico-chemical parameters are analysed: water and air temperatures, visible waste material, odour, colour, turbidity, pH, conductivity, dissolved oxygen, BOD5, COD (from $KMnO_4$), suspended solids, volatile solids, alkalinity, hardness, Ca^{+2}, Mg^{+2}, Na^+, K^+, SO_4^{-2}, H_2S, Cl^-, Nh_4^+, NO_3^-, NO_2^-, phenols, Fe^{+2}.

Biological analyses are as follows: Saprobity index (Liman), productivity index, and Putee-Buck number.

Microbiological analyses are: MPN of bacterial coli per litre, total number of microorganisms per litre.

2.2 WATER RESOURCES

Of the total area of the FR of Yugoslavia, which amounts to 102,173 km^2, the Danube river basin covers about 88,919 km^2, or 87% of the state territory. Annual rainfall in the Danube river basin in the FR of Yugoslavia is about 74.0 km^3 on average; of this quantity about 23.5 km^3 runs off and the remainder of about 50.5 km^3 accounts for evapo-transpiration. There is also an annual inflow of about 154.5 km^3 in these regions, so that the total annual run-off of the Danube, at the exit from the FR of Yugoslavia is about 178 km^3. The inequality of all basic components of the hydrological balances is very high as regards time and space.

Annual rainfall is the lowest in the north of the country, amounting up to 500 mm on average, and the highest in the south-west, with over 4,500 mm on average, but this water runs off into the Adriatic Sea. During the vegetation period, rainfall in some regions is only about 28% of the annual average. It is thus necessary to supply agribusiness with additional water quantities by irrigation. In other regions the amount of precipitation reaches 60%. The characteristics of the flow of the Danube at its point of entry into the FR of Yugoslavia and in the cross-section on the border with Romania, as well as at the confluences of direct tributaries are given in Table 1.

This Table shows a high ratio between minimal (Qmin 95%) and average (Qav) flow rates especially on smaller watercourses, where it ranges from 15 to 117.5 times. This provides ample evidence of the torrential nature of some watercourses and a need to protect the surrounding land and settlements from flood. At the same time measures for flow balancing are needed, to provide more effective utilisation of natural surface water resources.

TABLE 1. Characteristic flows of the Danube river and its tributaries.

River	Profile	Area (km^2)	Q$_{min}$ 95% (m^3/s)	Q$_{av}$ (m^3/s)	q$_{min}$ (l/s/km^2)	q$_{sr}$ (l/s km^2)
Danube	Bezdan	210,250	837.0	2,263	4.0	10.8
Danube	V. Gradiste	570,375	1,800.0	5,466	3.2	9.6
Tisa	at the mouth	148,973	126.0	794	0.8	5.3
Timis	at the mouth	10,280	0.4	47	0.04	4.6
Sava	at the mouth	95,132	287.0	1,570	3.0	16.5
Morava	at the mouth	38,345	35.0	232	0.9	6.1
Mlava	at the mouth	1,886	0.7	12	0.4	6.4
Pek	at the mouth	1,233	0.6	9	0.5	7.3
Timok	at the mouth	4,510	1.2	31	0.3	6.9

TABLE 2. Characteristic flows of the rivers in the Republic of Montenegro.

River	Area (km^2)	River Length (km)	Precipitation (mm)	Qav (m^3/s)	Runoff q$_{av}$ (l/s.km^2)	Runoff Coeff. (-)
Zeta	1,597	85	2,376	100.0	62.8	0.832
Morača	3,270	102	2,332	201.0	62.0	0.838
Piva	1,784	94	1,837	77.0	43.7	0.741
Tara	2,040	147	1,628	80.9	40.1	0.771
Ćehotina	810	77	895	14.2	17.5	0.618
Lim	2,875	123	1,235	77.5	27.4	0.706
Ibar	414	34	1,061	6.8	16.4	0.487

Approximately 70% of the drinking water in Yugoslavia is abstracted from groundwater sources and therefore particular attention is paid to their control. The large number of ground water resources are closely connected and influenced by watercourses.

3. Waste Water Discharge In Yugoslavia

Ninety per cent of wastewater produced in the FR of Yugoslavia is discharged into the Danube and its tributaries. The quantities of wastewater from the major sources and their present quality are given in the subsequent sections.

3.1 MUNICIPAL DISCHARGE

Municipal wastewater is treated in a relatively small number of settlements, and mostly in settlements that are located near smaller watercourses. In the FR of Yugoslavia, municipal wastewater is treated in more than 40 urban settlements, mostly by biological treatment. The efficiency of treatment varies: in the case of mechanical treatment it is up to 40% and in the case of biological treatment up to 95%, depending on the maintenance of equipment. In the Danube river basin, there are 37 plants for the treatment of municipal waste, their total capacity being c. 2,150,000 population equivalent (PE) (in Vojvodina c. 570,000 PE and in central Serbia 1,578,000 PE). Additionally, there is a wastewater treatment plant in the city of Podgorica (Montenegro) with a capacity of 50,000 PE in the Adriatic Sea watershed.

In addition to these central municipal treatment plants, there are treatment plants in some parts of urban settlements, in major tourist centres and facilities, as well as in weekend cottage complexes. These are mostly small, biological treatment plants with a capacity of up to 500 PE.

The quantity of municipal waste discharged in 1997 is estimated at a level of 1,056 x 10^6 m³/year for the FR of Yugoslavia. The predominant part (926 x 10^6m³/year) is discharged in the Danube watershed in the FR of Yugoslavia. About one third of municipal wastewater stems from the settlements located in the vicinity of the river Danube (c. 40%), while the largest quantities are produced in the catchment area of the Velika Morava River. This is logical in view of the number of inhabitants and industrial facilities in the settlements in this watershed area.

At the present time, about twenty municipal waste treatment plants, with a total capacity 2,000,000 PE, are under construction. The progress of these facilities varies from 10% to 60%. Design documentation is being prepared for about another twenty municipal waste treatment plants.

4. Industrial/Mining Waste Water Discharge

A number of minor industrial plants located in urban environments discharge wastewater into the sewerage. Larger industrial plants are most often located outside the settlements, usually on the banks of rivers or in their immediate vicinity, as are all mines. Waste from these facilities is discharged directly into watercourses and canals with or without advance treatment. Table 3 below shows the quantities of industry and mining wastes discharged directly into watercourses in 1997, in 10^6m³/year.

There is no estimate for shipping discharge quantities nor is it possible to make one, because such quantities depend on the engagement of the FR of Yugoslavia river fleet, as well as on the number and structure of foreign vessels in transit. By far the largest quantity of industry and mining wastewater is discharged into the Sava and its tributaries, due to a high specific consumption by the plants located in the river system. However, the catchment area of the Timok is especially affected by industrial and mining discharge, due to the composition of such water, the degree of treatment and the recipient's capacity.

In the Danube river basin there are about 120 industrial waste treatment plants, most of which provide only conventional or minimal treatment so as to meet the requirements for waste discharge into the sewerage. Only twenty or so larger industrial plants located on the banks of the Danube and its tributaries have the facilities for full treatment of wastewater, and some of these plants are only partly in operation. Ten industrial waste treatment plants are under construction and are more than half-ready. Design documentation is nearing completion for a further ten.

TABLE 3. Industry and mining wastewater discharge in 1997, in $10^6 m^3$/year.

Area - catchment	Industry and Mining
FR Yugoslavia	881
Danube (total in FRY)	731
Sava	290
Morava	180
Mlava	5
Pek	9
Timok	33
Tisa	95
Timis	5
Danube direct catchment	114

4.1 AGRICULTURAL DISCHARGE (MAJOR POINT SOURCES)

Cattle raising in the FR of Yugoslavia and the Danube River system is carried out mostly by the private sector. Cattle and pig raising is carried out mostly on co-operative farms and in enterprises, and all large farms (with more than 5,000 porkers or 500 bullcalves) are exclusively in social ownership.

In the Danube River basin there are 100 cattle farms, each of them raising 1,000 heads of cattle on average. They are less significant as point sources of water pollution due to a dry method of manure disposal. There are 130 pig farms, with about 1,200,000 porkers altogether, and they represent the main point sources of surface and ground water pollution in the Danube River system and especially in Vojvodina (the catchment areas of Tisa, Timis and Sava).

On the farms with a capacity of up to 20,000 porkers, a combination of dry and wet manure disposal is used, while on the farms with over 20,000 porkers, only the dry method is used. There are a total of 43 farms with a capacity of 10,000 or more porkers, 34 of which are located in Vojvodina.

It is held that the organic load of one head of cattle is equivalent to the load of 30 people and the load of one pig is equivalent to 5 persons. It can thus be concluded that the organic load on farms in the Danube river basin in the FR of Yugoslavia amounts to 9,000,000 PE. There is no doubt that a part of this load enters surface and ground waters.

Wastewater is discharged most often into lagoons or natural depressions, and after being stored for about 6 months is used for fertilising agricultural land. Only a very small number of farms have facilities for technological waste treatment (aerators, separators, biological gas production) and their functioning is problematic. Part of wastewater from farms penetrates into ground water by seepage through the soil; thus, the contamination of watercourses, hydro-reclamation canals and impounding reservoirs is not rare. The farms that are located in the vicinity of sources of water supply or recreational zones pose a special hazard.

There are no precise data on the quantities of wastewater discharged from farms into lagoons, nor is it possible to estimate the quantity of such water entering rivers and canals, since this occurs only in incident situations (when the lagoons are prepared for emptying, or if a storm is accompanied by abundant rainfall).

4.2 MUNICIPAL SOLID WASTE DISPOSAL

The problem of municipal waste management in the FR of Yugoslavia is very serious because, at present, there is no waste disposal area that conforms to sanitary criteria (Yugoslav or international) with respect to the selection of site, construction and method of use. Nor is there any primary selection or separation of secondary raw materials, so municipal waste contains not only conventional domestic waste but also toxic matter.

Municipal waste is most often disposed of in trash dumps (45%), which do not even meet basic sanitary criteria, and which may be already full or need to be closed down right away. Of the total capacity, 32% of the existing trash dumps could be used for another 5 years, and 20% of them even longer, if they were reconstructed in accordance with sanitary criteria and legal provisions. Only 3% of waste disposal areas satisfy basic environmental protection criteria.

80% of existing trash dumps are located in the immediate vicinity of watercourses and, at times, on their very banks. Bearing in mind that the soil is alluvial and that no measures have been taken to prevent the seepage filtrate from penetrating into the soil, the result is permanent direct or indirect soil and ground water contamination. In some municipalities, in periods of high water, torrents carry trash down the watercourse, so that the consequences are apparent over a much broader area than simply the actual waste disposal area.

A special problem is posed by the fact that some waste disposal areas are located on the banks of the rivers or impounding reservoirs which are sources of water supply, thus creating an actual hazard from the penetration of pollutants.

In addition to municipal waste, waste from industrial plants is also disposed of in the majority of these waste disposal areas. Thus, apart from large organic pollution, the seepage filtrate also contains toxic matter.

The situation in rural areas is identical, since the depressions, ravines and banks of minor torrential watercourses are used for uncontrolled trash and other waste disposal.

This method of trash dumping, as well as the use of mineral fertilisers and manure has resulted in bacteriological and chemical contamination of the first water-bearing stratum, so that in the whole of Vojvodina and in a large part of central Serbia and Kosmet this stratum cannot be used for domestic water supply.

4.3 HAZARDOUS SOLID WASTE DISPOSAL

Hazardous waste produced in industrial plants is temporarily disposed of in the plant itself, but very often inadequately. Part of this waste is liquid and is stored in metal and plastic barrels, tanks and concrete pools. Solid waste is stored in plastic bags on concrete bases, often without a shed, or in improvised cassettes. The basic criteria of safe disposal are satisfied only by a small number of temporary storage facilities.

According to the preliminary hazardous waste register, in the FR of Yugoslavia about 224,000 tons of hazardous waste (as specified by the Basle Convention) is produced each year. An almost equal quantity of waste is also produced, for which a hazardous or non-hazardous classification has yet to be determined.

So far, there have been several incidents of penetration of industrial hazardous matters into watercourses, which have caused pestilence of hydrobions in the polluted watercourses and the suspension of water abstraction for the water supply of the settlements situated downstream from where the accident occurred.

5. Discussion and Conclusions

Most of the regulations inherited from the former Yugoslavia are based on recipient standards. In the R. of Montenegro effluent standards were introduced through new regulations adopted two years ago. In relevant strategic documents concerning long-term water economy development in the R. of Serbia and the R. of Montenegro, effluent standards and a combined effluent-recipient approach have been adopted (Draft of Water Economics Plan of Republic of Serbia and Draft of Water Economics Plan of Montenegro).

Water quality monitoring in the FR of Yugoslavia is done by the hydrometeorologic institutes of the R. of Serbia and the R. of Montenegro, and at the present time seem to be of a satisfactory standard.

Municipal and industrial wastewater treatment in Yugoslavia is not of an adequate standard. This is also the case with municipal and industrial solid waste disposal. Although some regulations concerning municipal and industrial waste have improved in the past few years, the implementation of new regulations and appropriate waste treatment, disposal and reuse is hindered by the difficult economic situation and the fact that the country is going through a transitional period.

Basic preconditions for the future development of water pollution control in the FR of Yugoslavia are the economic development of the country, and its joining the international financial flows. The fulfilment of these preconditions will allow a great number of strategic objectives in the field of water pollution control to be achieved in the near future.

6. References

1. Annual Report on the Quality of Surface Waters 1991-1997, City Institute of Public Health, Belgrade, 1997.
2. Hydrologic Yearbooks on Water Quality 1991-1997, Republican Hydrometeorological Institute of R. of Serbia, Belgrade, 1997.
3. Hydrologic Yearbooks on Water Quality 1991-1997, Republican Hydrometeorological Institute of R. of Montenegro, Podgorica, 1997.
4. Physical Plan of the Republic of Serbia, Ministry for City Planning, Housing, Municipal Services and Construction of the RS, Institute for Architecture and City Planning of Serbia, Belgrade, 1996.
5. Water Economics Plan of Republic of Serbia (final draft), Institute for Water Resources Management "Jaroslav Černi", Belgrade, 1997.
6. Water Economics Plan of the Republic of Montenegro (working material), Institute for Water Resources Management "Jaroslav Černi", Belgrade, 1998.
7. Ljubisavljević D. et al. Legal aspects of Municipal Hydraulic Engineering (in Serbian language), Nauka, Beograd, 1997.

Part IV: Support for a Balkan Water Network

EUROWATERNET- A FRESHWATER MONITORING AND REPORTING NETWORK FOR ALL EUROPEAN COUNTRIES

T. LACK
European Topic Centre on Inland Waters
Water Research Centre
Medmenham, Marlow, Bucks SL7 2HD, U.K.

1. Introduction

European Union (EU) Member States have monitoring networks in place to assess inland water quality (essentially to determine the state and trends in the physico-chemical and biological quality of rivers, lakes and groundwater) according to their national or international/European requirements.

Information provided by countries to the European Commission may seem to be an important source of data for the European Environment Agency's (EEA) needs, but inspection reveals a great disparity in its nature and comparability. The information required by the Commission is primarily for assessing implementation of and compliance with directives, rather than for assessing the status of and temporal changes in water resources.

Therefore, to fulfil its legal mandate (provision of objective, reliable and comparable information enabling the Commission and Member States to assess, frame, implement, further develop or modify European environmental policy) the EEA has designed a European monitoring and observation network for inland waters: EUROWATERNET, which is based almost entirely on existing national monitoring networks.

In parallel, the proposed Water Framework Directive (WFD) will require Member States to monitor the status of surface water and groundwater at the catchment and sub-catchment level. Technical specifications of this monitoring network are being discussed with the Council Presidency, the European Commission (DGXI), Member States experts and the EEA and its Topic Centre on Inland Waters (ETC/IW). Emphasis is placed upon the necessity to obtain comparable information on quality assessments between Member States, and it is likely that EUROWATERNET will provide a suitable mechanism.

This paper describes the EEA's data and information needs and the water monitoring requirements arising from them, with particular stress on EUROWATERNET.

J. Ganoulis et al. (eds.), Transboundary Water Resources in the Balkans, 185–191.
© 2000 *Kluwer Academic Publishers.*

2. Role and Information Needs of the EEA

2.1 THE MANDATE OF THE EEA

- The EEA was set up by Order of the Council of Ministers in 1990 and has a number of statutory duties, chief of which is to provide the European Union and Member States with:

'objective, reliable and comparable information at a European level, enabling them to take the requisite measures to protect the environment, to assess the results of such measures and to ensure that the public is properly informed about the state of the environment'.

In the field of water, the European Topic Centre on Inland Waters was appointed in December 1994 to act as a centre of expertise for use by the Agency and to undertake part of the EEA's multi-annual work programme, in particular that relating to:

- The assessment of status and trends of surface and ground water quality and quantity.
- How that relates and responds to pressures on the environment (cause-effect relationships).

2.2 INFORMATION NEEDS OF THE EEA

Information is therefore needed on the status of Europe's water resources in order to detect changes over time (trends), and how the status and trends are related to the pressures on the environment and to policies directed at the pressures or the societal forces creating those pressures. This information will only be credible if its component data are comparable and give a representative assessment of water types within a Member State and across Europe. A process of Integrated Environmental Assessment (IEA) has been adopted by the EEA, which operates within a framework of Driving forces, Pressures, State of environment, Impact on environment, and Responses in the form of policy and regulations (DPSIR).

Member Countries design networks and monitor water resources according to their national requirements (legal and operational) and international obligations (e.g. EU Directives and International Conventions).

Information from Member States' reports to the Commission might appear to be a rich source of information for the EEA's needs, but inspection has revealed that, by and large, the reports are not suitable mainly because:

- The degree of comparability depends on the interpretation of the designation rules of directives, and national differences in how these are implemented.
- Data requirements for those directives that require routine monitoring are generally site specific, either at sites designated for a specific use and/or at sites affected by a specific discharge.

The Exchange of Information Decisions requires data from agreed sites in main rivers. The choice of sampling location is, for most directives, related to areas designated in an ad hoc way by the Member States. It is not surprising therefore that a comparison of quality across Europe based on these designated waters gives an incomplete and heterogeneous picture.

The same observation has been made for information obtained from International Conventions, which represent only the main transboundary catchments in Europe.

Consequently the EEA has designed a European Water Monitoring Information and Reporting Network to provide comparable and representative information on the state and pressures on the environment through harmonised methodologies of selection and monitoring of all water bodies. EUROWATERNET is the process or system based upon the European Environmental Information and Observation Network (EIONET). It is a network of people, organisations, hardware and software, whereby data can be gathered from within Member Countries. This allows the EEA to obtain the information it requires to fulfil its legal tasks and to enable the Commission and Member Countries to evaluate the effectiveness of policy. It should be emphasised that the information provided by the network will not be for the assessment of compliance of Member States with the requirements of European Commission Directives, a responsibility held by the Commission.

3. EUROWATERNET

3.1 DEFINITION AND GENERAL CONCEPT

EUROWATERNET is the process by which the EEA obtains the information on water resources (quality and quantity) it needs to answer questions raised by its customers. Questions may relate to statements on general status (of rivers, lakes and groundwater) or specific issues (e.g. water stress, nutrient status and acidification at a European level.

The key concepts of EUROWATERNET are:

- It samples existing national monitoring and information databases.
- It compares like-with-like.
- It has a statistically stratified design 'tailor-made' for specific issues and questions.
- Its power and precision are known.

The network is designed to give a representative assessment of water types and variations in human pressures within a Member Country and also across the EEA area. It will ensure that similar types of water body are compared. The need to compare like-with-like is achieved with a stratified design with the identified and defined strata containing similar water bodies. The use of the same criteria for selecting strata and water types across Member Countries will help to ensure that valid status comparisons will be obtained.

A basic network of river stations and lakes based on the relative surface area of countries is the first step for Member Countries. However, it is likely that these will not answer all the questions raised by the EEA's customers, or perhaps not with the desired precision and confidence.

Therefore, a flexible approach will also be required for the selection of other monitoring stations included in national networks, in order to be able to answer more specific questions, such as "what is the extent of acidification in Europe?" or "what is/will be the impact of the Urban Waste Water Treatment Directive on water quality?" This is because the stations required for these questions may not always be located on the same water bodies/catchments as the basic network stations. In addition, more specific and detailed pressure information might be required. Thus to meet some of the EEA's information needs, site selection within each country must be issue or question driven. This, if necessary (in the light of experience), will form the impact network of EUROWATERNET.

A network fully representative of the differences in, and variability of quality, quantity and pressures found in all water body types across Europe, would be expected to answer most questions asked of the EEA. It is the long-term aim to make EUROWATERNET fully statistically representative. This will be achieved through the experience gained in implementing the basic and impact networks. This development will need to take into account the number of stations required to answer questions with defined, or at least known, levels of precision and confidence, and with knowledge of any inherent bias (for example, towards the most polluted water bodies) in the selected river stations, lakes or groundwater sampling wells.

3.2 FUNCTIONALITY OF EUROWATERNET

EUROWATERNET has to be able to answer questions about the state of the environment and a number of "status indicators" have been specified for that purpose. These indicators will also form part of the annual national reporting process, so there is now a real possibility of single sources of national data being used for multiple purposes (e.g. EEA, Commission, OECD, EUROSTAT and so on).

EUROWATERNET also has to be able to answer questions regarding the pressures from society on the water environment, to be able to relate to the changes in state. Examples of the physical characteristics and pressure indicators are shown in Table 1. Pressure indicators are of greatest value if they are expressed on a catchment or a sub-catchment basis. This means that EUROWATERNET will be encouraging the use of GIS in displaying both state and pressure information. This will be in harmony with several Member Countries' use of river catchments as the basic unit of management, and is also supportive of the Water Framework Directive that takes a similar approach.

3.3 DETAILED TECHNICAL SPECIFICATIONS

It is inappropriate to provide here the detailed technical specifications of EUROWATERNET. They have been developed by the European Topic Centre on Inland Waters and thoroughly discussed and agreed by Member Countries at an EEA workshop held in Madrid in 1996. The technical specifications have been published by the EEA. (European Freshwater Monitoring Network Design, Ed. S C Nixon, Topic Report 10/1996, EEA Copenhagen 1997) and EUROWATERNET. Technical Guidelines for Implementation, EEA Technical Report No 7, June 1998.

TABLE 1. Examples of physical characteristics and pressure information required for each river station and lake in the basic network, and the groundwater network

	rivers	lakes	ground water
Physical characteristics			
- depth (mean)		✓	
- surface area		✓	
- catchment area upstream of station/lake	✓	✓	
- catchment area recharging/affecting groundwater body			✓
- station/lake altitude	✓	✓	
- longitude/ latitude	✓	✓	✓
- upstream river length to source	✓		
- hydrogeology			✓
- aquifer type			✓
- aquifer area			✓
- soil type/geology of catchment	✓	✓	✓
Pressure information			
- population density in (upstream) catchment	✓	✓	✓
Upstream catchment land use such as:-			
- % agricultural land	✓	✓	✓
- % arable	✓	✓	✓
- % pasture land	✓	✓	✓
- % forest	✓	✓	✓
- % urbanisation	✓	✓	✓
Point source loads entering upstream	✓	✓	✓
Fertiliser usage in catchment upstream	✓	✓	✓

4. Relationship with the Proposed Water Resources Framework Directive

The Water Framework Directive (WFD) addresses all qualitative and quantitative aspects of surface and ground waters. It is stated in the Explanatory Memorandum "that all measures to achieve the environmental objectives for a sustainable protection and use of water are co-ordinated and their effect overseen and monitored within river basin". The objective is to achieve for all waters a "good" quality status and to prevent deterioration of current quality.

In this context, the WFD will require Member states to establish monitoring networks enabling them to:

- Assess and report on water status (chemical and ecological status for surface waters and chemical and quantitative status for groundwater) using a harmonised classification scheme.
- Diagnose problems, chart progress and develop appropriate action programmes for protection and improvement of waters.

Considering the WFD's importance to and impact on Member States, and the need to ensure that Community policy is applied in a coherent and consistent way, it is essential that countries implement consistent monitoring programmes and harmonised interpretation methods to guarantee the implementation of coherent action programmes across Europe.

The technical specifications of monitoring required under the WFD are currently being developed by the Commission, the EEA and Member States. The explanatory memorandum of the Directive (COM(97)49 final) states that "data is collected largely for operational reasons to inform decision making within the individual river basins." The articulation with EEA monitoring is also mentioned as follows: "the monitoring programmes required by this proposal will extend the range of stations which the Agency can draw upon in developing its network". It seems very likely therefore that EUROWATERNET will be able to satisfy most of the monitoring needs of the WFD. Work is currently in progress to test the feasibility.

5. Benefits to Member States, Progress to Date and the Way Forward

5.1 BENEFITS TO MEMBER STATES

Additional monitoring is expensive and is unlikely to be undertaken purely for the "European need." This is why EUROWATERNET is firmly based on existing national programmes. By and large most countries have an adequate national monitoring network for rivers to meet the EEA's needs (in terms of number of stations, frequency of monitoring and determinants monitored). This is less true with regard to groundwater and even less true for lakes/reservoirs, although there are some exceptions.

Where national networks can be shown to be insufficient in terms of providing data to allow a representative European picture to be obtained, it is also likely that a case can be made to the Member Country concerned, showing that the country itself is not obtaining a representative picture of national water resources. An example of this could be if a national programme has been designed to provide data on impacted lowland rivers. It would be safe to say therefore that this programme will not tell anything about unpolluted upland streams. If there were no other national programme on upland streams, then when the European picture on upland streams was painted, for that country the picture would have to be blank. However, this situation, should it arise, could be mitigated if there were, within the country, *a priori* knowledge about the state of upland streams and the stability of the pressures upon them, so that a decision not to monitor them (or monitor them less frequently) could be justified. Therefore the decision whether or not to undertake additional monitoring remains within national control. Member Countries also have the freedom to decide if and when to discontinue or reduce national programmes.

As stated elsewhere in this paper, there is not enough comparable information to obtain a quantitative assessment of water resources across Europe at present. This can lead to unfair or incomplete comparisons being made and wrong conclusions drawn about the effectiveness or otherwise of policy instruments, which have often been implemented at great cost to Member Countries.

Therefore, Member Countries will benefit from:

- A better knowledge of policy impact leading to improved European policy, as a result of the better information.
- Providing inputs that show their country in the best (correct) light, making full use of national data.
- Providing information on international environmental issues in a consistent and standardised way.
- Informing on policy initiatives by providing comparative information on performance in other countries.
- Providing their citizens with information on the state of the environment in their own country, neighbouring countries, European regions and the whole of Europe.

5.2 PROGRESS TO DATE AND THE WAY FORWARD

The overall objective is to have EUROWATERNET fully operational on the EIONET by December 2000. Currently, data obtained through the Eurowaternet process has been provided by around 12 of the 18 EEA countries and by 4 of the Phare countries. Commitment to implement EUROWATERNET has been given by all member countries of the EEA (31 at present).

MONITORING WATER SYSTEMS: THE USA EXPERIENCE

D. G. FONTANE
Department of Civil Engineering
International School for Water Resources
Colorado State University
Fort Collins, CO 80523 USA

1. Introduction

World-wide, the management of water systems is becoming more complex. There are numerous reasons for this including increasing population and urbanisation. Both population and urbanisation will continue to rise in the foreseeable future. Therefore, the degree of complexity will continue to increase. Djordjevic (1993) identified three indicators of this complexity: increasing difficulties in providing the required quantities of water for different purposes, increased problems associated with hydrologic phenomena such as floods and droughts, and increasing danger to humans and the environment from water pollution and destruction of water-related ecosystems. Water managers rely on good quality data and information to guide their decisions. As Djordjevic [1] so succinctly states, *"Cybernetics has introduced the new element,* **information**, *without which a survival of organised large systems would not be feasible."*

Monitoring is the process that provides this required data and information. As the nature of the water resources system evolves, the nature of the monitoring system must evolve also. The physical infrastructure for water management is increasing in response to rising demands. In addition, the decision process is becoming more complex as more and more stakeholders are being brought into the decision process, and water-related regulations and legislation continue to expand. Water managers must constantly reassess their information needs. Further, as Grigg [2] points out, water managers are only one group that relies upon water resources data and information. Other groups include: regulatory agencies, environmental organisations, elected officials, educational organisations, recreational organisations, and the general public. Monitoring, therefore, must respond to the continually evolving information needs of diverse communities. Finally, monitoring systems must also respond to the rapid changes in information technologies. Computing software and hardware has radically changed the nature of information production and distribution.

In the design of a monitoring system, three questions must be addressed: what to monitor, how to monitor it, and how to provide the monitored information? Fontane [3] categorised these issues into "Availability, Accessibility, and Analysis" and they are illustrated in Figure 1. Availability refers to the ability to effectively and efficiently

J. Ganoulis et al. (eds.), Transboundary Water Resources in the Balkans, 193–201.

measure the parameter of interest. Some water resources information is much harder to collect than others. This difficulty occurs for both physical and social reasons. Consider a commonly available parameter of interest such as water flow rate. Control points on a river can be identified to monitor river stage and to develop stage-discharge relationships. Similarly, discharge versus percent-open relationships for valve and gates can be established. As a result, good records have been developed for river flow and reservoir releases. However, it has been rare to develop good measuring technologies to monitor urban and agricultural water use or demand. Some of the difficulties have been technological, such as the problem of developing effective monitoring of leakage in an urban water supply network. Yet many of the difficulties have been more social in nature. It is important to recognise that human interaction has traditionally been required to observe and record the information. Therefore, the motivation to dutifully collect and record the information depends upon the perceived value of that information to the human observer. Additionally the cost of collection of such information is high.

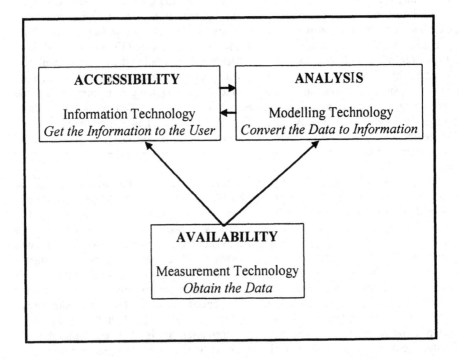

Figure 1. Three aspects of information monitoring systems.

The reliance on the human observer/recorder has lessened significantly in recent years with the development of effective remote sensing technologies. The provision of the automated data observer/recorder has essentially produced real time availability of large quantities of information. Concurrently, this has produced a need to store and manage an enormous quantity of information. There are still major gaps in information that are difficult to obtain for technological and other reasons. For example, ecological sampling and some water quality parameters measurements are still human labour intensive. However, it seems that the future will yield increased availability of a wide range of water-related information.

The issue of "accessibility" deals with the ability to acquire the measured information. There are two major aspects of this issue. The first has to do with the physical characteristics of the recorded data. Prior to digital technologies, data was stored in written form. The analogue strip charts from the river stage recorders were processed into digital values that were then recorded in written form and stored in physical locations such as file drawers. This process required a certain amount of time. Therefore, real time information was extremely scarce. Certain types of data were published in report or book form (certainly a time consuming process) to make the data generally available. An example in the USA would be the published flow records of the US Geological Survey. This data was then "accessible" via the report to the water management community. However, in order to use this data for analysis with computers, it had to be coded into digital form. This step required time, again negating the concept of real time information, not to mention the potential loss of information in the multiple processing required of the data. Again, the development of remote sensing technologies has changed the accessibility issue greatly. Data is sensed, recorded in digital form, and transmitted in digital form to a database where it is directly available in real-time to the collecting agency. Technology also allows this data to be automatically posted to the Internet. If the collecting agency chooses to put its real-time information on the Internet, it is immediately available to the world.

The second aspect of accessibility has to do with the organisational control of the information. The information might exist in digital format, but it might not be easily accessible to anyone outside of that organisation. Reasons for restricted information sharing might include security concerns involving interagency, interstate, or international transboundary conflicts. Another reason could be policy directives that require the information can only be provided if the user pays for it. Finally, the authorised mission for an agency might be simply to collect and record the information. Providing information to other agencies may not be established as a primary role for the organisation. The information may be obtainable, but the process to get the data may take weeks or even months. The technology of the Internet, in many cases, has revolutionised the issue of accessibility of information.

The issue of appropriate analysis tools has aspects of both information handling technologies (databases) and information processing technologies (models and spreadsheets). In the 1970s, database hardware and software technologies were primitive, compared to current standards, and the amount of information that could be processed was limited. Also, access to databases could be quite slow. Water resources models and spreadsheets were likewise limited by computing technologies. The personal computer revolution of the 1980s led to a great expansion of the use of models by water resource engineers and the development of spreadsheet technology. However, in the 1990s, the enormous increases in computing hardware capabilities and reliable networking technologies provided unparalleled analysis power on a desktop (or laptop) computer.

2. Monitoring in the USA

2.1 FEDERAL LEVEL PROGRAMMES

Monitoring of water systems in the United States of America (USA) has historically been centralised at the Federal (national) level. The primary water resources monitoring agency at the Federal level is the U.S. Geological Survey (USGS). Other important agencies with monitoring components that support water resources management are the National Weather Service within the National Oceanic and Atmospheric Administration (NOAA) and the U.S. Environmental Protection Agency (EPA).

The USGS began operating its first streamgauging network in 1889. Initially the focus of the monitoring activities by the USGS was on appraising the water resources and therefore flow quantity was the primary parameter measured. During the 1970s passage of environmental legislation, such as the Clean Water Act and Endangered Species Act, served to broaden the monitored parameters to include water quality and other environmental parameters. The agency's 1998 report to the U.S. Congress details the current status of its water resources monitoring activities [4].

2.2 MAJOR GOALS

The USGS 1998 report describes five major goals for the monitoring system: interstate and international transfers, water budgets, flooding, water quality, and long-term changes. These goals and their users are illustrated in Figure 2. The first goal for the monitoring system is to provide *neutral* data for use in allocation of interstate waters. The stream gauging network is operated by the USGS, yet funding is contributed by both the USGS and other Federal, State and local water agencies. For the fiscal year 1998, the total funding for the USGS streamgauging network was $89 million (US) of which $28 million came from other Federal agencies and $35 million came from State and local agencies. Therefore, the system is operated for the benefit of all contributing partners. This arrangement provides the motivation to consider the needs of all the partners and to insure the accuracy and neutrality of the data. River reaches that cross an interstate or international boundary and have at least a 500 square mile drainage area at the crossing point have been identified. Currently, approximately 56% of 331 identified sites are adequately gauged. The selected gauged sites were identified as the most important among all of the sites.

River basins represent an important surface water hydrologic boundary. Although they seldom correspond with political boundaries, information on a river basin scale is very useful for water policy considerations at a national level. The second monitoring goal is to develop information for water budgets. Specifically, the objective is to quantify the outflows of the major river basins in the USA. Approximately 77% of the 329 identified major river basins are adequately gauged.

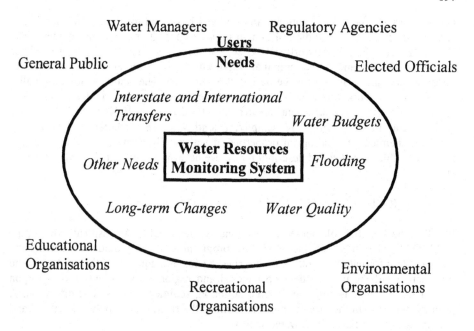

Figure 2. Illustration of the users and needs for the monitoring system.

The third goal for the monitoring system is to provide streamflow information to flood management agencies and the general population at risk from flooding. Both long term and real time information is needed. Long term monitoring is necessary to identify flood characteristics and flood zone boundaries. It is needed for the design of flood management infrastructure. Real-time information is needed for operation of flood management infrastructure and for flood warning systems. The National Weather Service (NWS) is the Federal agency responsible for developing and issuing flood forecasts and flood warnings. The USGS streamgauging system provides most of the streamflow information used by the NWS. The NWS provides forecasts to approximately 3,000 river locations in the conterminous United States. With increasing urbanisation in flood prone areas, the potential for flood loss increases. Therefore, the importance of monitoring for flooding situations will grow in the future.

The fourth monitoring goal deals with the water quality category. This goal involves providing information to guide the improvement of water quality in watersheds with a specified level of water quality degradation The EPA [5] defined 677 out of 2,079 watersheds as having water quality degradation. Water quality degradation was present if at least 50% of the assessed rivers in the watershed failed to meet designated uses. The USGS currently provides adequate streamflow data on approximately 85% of these watersheds. During the past 30 years, the USGS has operated two national stream water quality networks. The purpose of these networks was to provide the status and trends of water quality on a national and regional basis and to provide information to improve scientific understanding of human impacts upon the natural environment. The Hydrologic Benchmark Network (HBN) consists of 63 watersheds. These watersheds are fairly small and have minimal environmental disturbance. Water quality measurements are available for the period 1962 to 1995. The National Stream Quality Accounting Network (NASQAN) covers selected watersheds throughout the USA. Stream water quality measurements are available for the period 1973 to 1995. The water quality data includes uniformly sampled (monthly to semi-annually) data for 63 physical, chemical and biological properties of waters. These measured properties are shown in Table 1. A detailed description of these networks is provided in Alexander, et al. [6].

2.3 WATER QUALITY ASSESSMENT PROGRAMME

In 1991, the USGS implemented the National Water Quality Assessment Programme (NAWQA). The intent of the NAWQA programme is to provide consistent and comparable information for 60 important river basins and aquifers in the USA [7]. The information will be used to develop national and regional assessments focused upon critical issues, such as non-point source pollution, sedimentation, and acidification. A primary objective of the programme is to identify the relationships between natural and human factors that most affect water quality.

The fifth goal is to monitor for long term changes. The intent is to identify long term trends in stream flow in representative streams within the USA. The nation has been divided into eco-regions with similar landscape, elevation, climate, and land use. A representative stream is one without significant regulation and diversions and it must lie entirely within the eco-region. The purpose of the collected information is to identify changes in stream flow resulting from climate changes, changes in land use, or changes in groundwater withdrawals. Currently, about 76% of the eco-regions have adequate gauging coverage.

Overall, the availability of water management information developed by the USGS is quite good. The accessibility of information is excellent, since the USGS has focused on applying satellite telemetry and the use of the Internet. The number of streamgauging stations with satellite telemetry rose from 14% in 1983 to 64% by 1997. One of the guiding principles of the USGS monitoring system is that the data should be freely available to its funding partners and the general public. Currently, there are approximately 4,000 streamgauging stations where near real-time data is posted to the Internet. This data can be accessed free by the public, as can certain historical data sets. Water quantity and quality data sets are included, as are spatial data sets suitable for Geographical Information Systems (GIS) software. Large data sets are available on CD-ROM for a nominal charge. Finally, tools for analysis of the information continue to increase. As an example, the USGS National Research Programme is developing improved watershed modelling technologies, including the Watershed Modular Modelling System with a GIS graphical user interface [8]. These products should ultimately be accessible via the Internet.

TABLE 1. Water properties and sample frequencies (adapted from Alexander, et al. (1996).

Water Properties	Samples per Year
Physical/field measurements: Temperature, specific conductance (field and lab), dissolved oxygen, pH (field and laboratory), suspended sediment, turbidity, instantaneous Stream flow.	4 to 6
Stream flow, daily mean	365
Major ions: Calcium, chloride, magnesium, potassium, silica, sodium, sulfate, fluoride, dissolved solids, hardness, alkalinity, bicarbonate, and carbonate	4 to 12
Nutrients and carbon (dissolved and total): Ammonia+organic (Kjeldahl) nitrogen, ammonia, nitrite, nitrate+nitrite, orthophosphate, phosphorus, and nitrogen	4 to 12
Organic carbon (includes suspended also)	4 to 8
Radiochemicals (dissolved, suspended sediment): Gross alpha and beta, radium-226, tritium, uranium	1 or 2
Biological measurements: Faecal coliform and streptococci	4 to 12
Periphyton (chlorophyll A and B), periphyton biomass	4
Phytoplankton (chlorophyll A and B), Phytoplankton (total count), identification of predominant forms	7 to 12
Inorganic trace elements (dissolved, total): Aluminium, arsenic, barium, beryllium, boron, cadmium, chromium, cobalt, copper, iron, lead, lithium, manganese, mercury, molybdenum, nickel, selenium, silver, strontium, vanadium, and zinc	3 or 4

In addition to the monitoring done by the USGS, the EPA monitors quality within the water resources system, particularly with respect to drinking water concerns, and the National Oceanic and Atmospheric Administration monitors and distributes the meteorological information.

Both of these agencies use the Internet to provide the information to the user. For the most part, it is provided free of charge. Major Federal and State water management organisations also provide monitoring functions. As an example, the U.S. Bureau of Reclamation is responsible for the operation of 59 hydroelectric power plants in the Western United States. They have developed an Internet based hydropower programme report that details the features of their hydropower facilities. They also provide a power generation statistics database that gives current and past power generation records for their projects on either a monthly or yearly basis. The Bureau of Reclamation also provides information via the Internet on many of their non-hydroelectric facilities and the amount of information provided is rapidly increasing. Water management agencies have moved from using the Internet primarily as a public relations tool, to using the Internet as a primary vehicle to share operational data within the agency and with the general public.

3. Future Directions

The United States is fortunate to have a well-established and centralised water monitoring system. The use of a co-ordinated monitoring system, such as that maintained by the USGS, has served to minimise duplication of effort and provide neutral, consistent and high quality information. Undoubtedly, the need for information will increase into the future, and this will require increased costs. The USGS 1998 report raises concerns that Federal funding for its monitoring program has not kept pace with the increased need for information. There are also identified needs within the USGS to modernise its monitoring and information delivery systems. The situation is similar for other agencies with monitoring functions. Technologically, it is possible to measure more information, to measure it faster, to analyse it better, and to make it available to the public more quickly. This will require an increased commitment of both human and financial resources.

As our understanding of the complexity of water related ecosystems improves, more and more data and information (for example, biological indicators) will likely be required to effectively monitor these systems [9]. Much of this information may be difficult to remotely sense and, therefore, will require labour intensive sampling and analysis processes. Additionally, this data will have to be made available to the public in a systematic format.

Data presentation via the Internet has improved significantly during recent years. However, more effort will be needed to assure the completeness and integration of the information. It is imperative that the units, location, measurement procedures, time and spatial scales, data format, and other appropriate information are provided with the data. Formats for data, specifically data to be downloaded, need to be designed for modern database and spreadsheet software. Better cross-referencing of information sources is needed. As an example, the USGS Internet sites for streamflow data within a watershed provide links to the EPA information on that watershed. Water quality data might be provided on web pages that also provide links to sites with the water quality standards for that river, related water legislation, and information on water quality modelling studies for that location. Providing this level of integration of information will require the commitment of human resources and a strong level of communication and co-operation between water management agencies. Murphy [10] describes a concept for an Internet based site to provide integrated water resources information on an international river basin scale.

4. References

1. Djordjevic, B., 1993, *Cybernetics in Water Resources Management*, Water Resources Publications, Highlands Ranch, Colorado.
2. Grigg, N.S., 1996, *Water Resources Management: Principles, Regulations and Cases*, McGraw-Hill.
3. Fontane, D.G., 1997, "New Frontiers in Information Science," Keynote address presented at the Annual Meeting of the Brazilian Water Resources Association, Vitoria, Brazil, November 19, 1997.
4. U.S. Geological Survey, 1998, *A New Evaluation of the USGS Streamgauging Network*, A Report to Congress, U.S. Department of the Interior.
5. U.S. Environmental Protection Agency, 1998, *Introduction: Index of Watershed Indicators*.
6. Alexander, R.B., Slack, J.R., Ludtke, A.S., Fitzgerald, K.K., and Schertz, T.L., *Data from Selected U.S. Geological Survey National Stream Water-Quality Monitoring Networks (WQN)*, U.S. Geological Survey Digital Data Series DDS-37, based upon the U.S. Geological Survey Open-File Report 96-337.
7. Leahy, P.P., and Thompson, T.H., 1994, *The National Water-Quality Assessment Program*, U.S. Geological Survey Open-File Report 94-70.
8. G.H. Leavesley, S.L. Markstrom, M.S. Brewer, R.J. Viger, (1996), "The Modular Modelling System (MMS) -- The physical process modelling component of a database-centred decision support system for water and power management," *Water, Air, and Soil Pollution* 90, p. 303.
9. U.S. Geological Survey, 1999, *Status and Trends of the Nation's Biological Resources*, Volume 1, U.S. Department of the Interior.
10. Murphy, I.L. (1997), *The Danube: a River Basin in Transition*, Dordrecht: Kluwer.

MANAGING HUMAN RESOURCES FOR WATER MONITORING: A ROLE FOR THE INTERNATIONAL WATER ENVIRONMENT CENTRE FOR THE BALKANS

R. D. CASANOVA
University of Nice-Sophia Antipolis (UNSA)
France

1. Introduction

The most important common problem of our planet may well be the quality of its water. Simply saying so however, does not produce solutions. With respect to transboudary water resources management in the Balkans, the concern lies mainly with regional aspects of the issue, however local considerations are equally as important. Based on experience in the management of public water resources, this paper details and illustrates the complexities of the problem, and shows how even a local approach can give satisfaction to collective needs.

This paper also shows that water management requires a multidisciplinary approach with many specialists, meteorologists, geographers, geologists, hydrologists and hydraulic engineers, working together with lawyers, managers, land planners, politicians, and teachers. Problems related to water resource management link the disciplines of physical and natural sciences, sociology, economics, ecology, and development planning, among others. If the approach is global, solutions are necessarily local and should include urban and water resources management with appropriately trained personnel.

To solve an environmental problem a systemic and global approach is needed to define the problem, and a pragmatic and operational approach to achieve an established goal. This paper offers guidelines for the establishment and operation of an International Network of Water-Environment Centres for the Balkans (INWEB), by promoting training and professional tools to help achieve its goals. It suggests outlines for:

- A post graduate course (to be held in English or another international language) for students and teachers involved in water management policy.
- Related courses to be offered in local language for those involved in the monitoring and resolution of technical issues in water management.

J. Ganoulis et al. (eds.), Transboundary Water Resources in the Balkans, 203–212.

204

2. Training Proposal Requirements

An adequate training proposal requires linkages among water resource and environmental issues as they relate to the goals of sustainable development. A prerequisite of an optimal training programme for water resource management is that the true needs of the field and the desired level of knowledge be clearly specified. This requires attention being paid to the following topics:

- Quality of life should be part of the subject matter and include references to landscapes, aesthetics, cleanliness, security, and public health. The goals of sustainable development provide a paradigm which would include references to its philosophy, considerations regarding the preservation of life, respect for people, solidarity, humanism, general equilibrium, respect for nature, and the assessment and prevention of industrial and natural risks.
- A range of scientific disciplines should be included in determining adequate training goals, including ecology, zoology, botany, biosphere, ecosystems, atmosphere, geology, legal, economics, waste water treatment, hydrogeology and hydraulics.
- The satisfaction of collective needs should be explored through political science and economics including subjects such as individual and collective autonomy, individual freedom and community needs, natural heritage and civic responsibility, amongst others.

Thus, before studying an environmental problem and exploring the need for monitoring of transboundary water resources, the exact nature of common interests should be agreed upon. Every country needs to train people to manage water and the environment for current and future generations. To do this, it is firstly necessary to identify the true needs, the exact field of study and the level of knowledge desired.

2.1 METHODOLOGY

The first step in managing human resources necessary for water resource monitoring is to analyse and understanding the problem. Secondly an appreciation is needed of the community's, or company's intentions with respect to specific problems with water resources, water quality, health, waste treatment, energy, training, and so forth. What goals exist, and what policies could help achieve them? Efficiency requires a pragmatic approach. In this case a permanent forum of all the specialists involved in the project is needed. If it is necessary to train people to fulfil the goals, who should be trained and at what technical level? Should they be regular, full-time students or can some technicians and engineers be re-trained? What are company goals with respect to this question, and what is the market for training?

This review should enable a proposal for a pragmatic working programme to be developed, and provide a follow-up evaluation of the training programme.

In summary there are five main questions:

- What are the laws and rules in the area where the project is conceived?
- What is the market for trained people?
- What is at stake from an economic point of view?
- Where do we start?
- What are scientific and technical problems?

Creating an International Network of Water-Environment Centres for the Balkans (INWEB) would be the best way to train the people needed for monitoring water; as it would provide a focal point where forums could be organised for professionals working together, and would then lead to the creation of an electronic network.

3. An International Balkan Water Resource Training Programme

3.1 COURSE THEMES

The following course themes are suggested:

- Human activities and their environmental impacts
- Water science: resources, quality, transport, pollution, management
- Watershed hydrology, karsts landforms and applications
- Modelling of hydrology, hydro-informatics
- Business, companies and the environment
- Environmental standards and environmental management (ISO 14 000)
- Environmental laws
- Environmental risks assessment
- Waste treatment technologies, waste management
- Environmental audits
- Soils and water pollution
- Eco-toxicology
- GIS as a tool for urban or land planning

3.2 GENERAL APPROACH

The proposed project for applied environmental research and training suggests the need to identify the market for such training in universities, companies and government departments involved in policy making. During their training period students would work as if they were in a firm of experts i.e. several students would work on the same project, but each of them would be in charge of a different part of it, according to his/her own speciality, the needs of the study and the student's country of origin. The training course would include one semester devoted to courses at the university and a semester during which the students would be trainees in a company or government department.

The proposed International School for Water Monitoring in the Balkans would adopt the European Credit Transfer System (ECTS) and be associated with a network of

206

Balkan universities. It would be possible later to exchange students, researchers and teachers as well, but the initial purpose would be to exchange training schemes in order to deliver credits (or diplomas) that would be valid for every country involved. For a regular training programme students could receive credits for a course after an examination according to the ECTS system (European Credit Transfer System). Courses could be completed within the Balkan countries or on the Web.

For suggested course outlines see the Annexes at the end of this paper.

4. Conclusions

This paper offers the following conclusions:

- The NATO ARW in Thessaloniki demonstrated the participants' mutual interest in transboundary water resources management in the Balkans, and the feasibility of their now working together to achieve common goals. The first step required is to create the tool for such regional co-operation i.e. to formally establish the International Network of Water-Environment Centres for the Balkans (INWEB).
- Once this has been achieved, then Balkan countries can work together to solve water management problems.
- One of these problems will be how to train, at low cost, water monitoring personnel.
- It will be difficult to organise all the professional training courses needed in the field of water monitoring in the Balkans. The first step, however, is to define the human resources needs of each country, so that courses and professional training can be organised with teachers from various countries, in accordance with the scheme presented below.

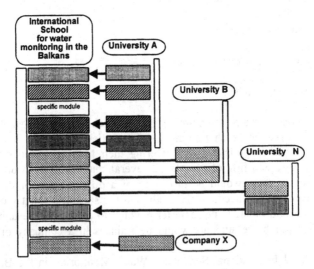

Figure 1. Definition of a training scheme.

5. Examples of Possible Postgraduate Level Courses for the Balkans

5.1 ANNEX I. INTEGRATED WATER RESOURCES MANAGEMENT – HYDRO-INFORMATICS

Water resources management today needs to integrate more and more different components, such as water quantity and quality, risks of pollution, scarcity of resource, environmental demands and users. The conflicting demands on the scarce resources of water call for an integrated approach to water resources management. Using examples from actual experience from case studies or other material the course could cover subjects such as:

• Concepts and tools in water resources management.
• River basin simulation.
• Environmental quality.
• Water use activities.
• Principles of hydrology.
• GIS technology.

The information revolution has fundamentally altered traditional approaches to design, planning and control of hydraulic, hydrological and environmental systems. As the capabilities of computers and communication networks have rapidly expanded, so have the complexity of simulation models and the means of acquiring, storing, retrieving and manipulating vast amounts of information. Its principle goal is to reduce the damage caused by human activity to acquatic ecosystems.

Thus the aim of the course is to introduce participants to hydro-informatics systems applied to different topics, such as urban water management or watershed management. The course combines lectures on the essential theory and the practice of hydro-informatics in several environments: storm and wastewater drainage and flood management. The course could use as material examples of urban areas, rivers and/or watersheds in different countries. These real life cases, supported by lectures and hydro-informatics, would demonstrate the scientific and technological methodologies compatible with integrated water management.

Course scheme (Duration: 12 hours)

Topics	Description
Hydro-informatics	Concepts and background
	Society and the market
	New technologies
Storm and wastewater drainage	Short and long term
	monitoring
	Simulation modelling
	Case study
River impact	Flow and quality processes
	Simulation modelling

5.2 ANNEX II. KARST LANDFORMS AND APPLICATIONS

Limestone covers 10% of lands. The action of water on limestone creates particular landforms called karsts. Usually considered as poor places they are in fact of economic importance. Karstic areas are important in the Balkan region.

Karsts are very attractive places for tourism (Guilin in China; Along Bay in Vietnam; Langkawi, Mulu and Niah in Malaysia, and the Adriatic coast), and include very important water resources. Karst springs are useful worldwide for water supply, but there is no filtering in karsts and water is usually contaminated. The study of hydrogeological basins of karstic springs has become increasingly important in order to prevent pollution.

There are often narrow gorges that are good places to built dams in karstic areas. However, as limestone contains caves, important leaks may exist. Some cases have been total failures. Karst areas are natural recorders where it is possible to find information about paleo-environments (seismic hazard, active faulting, past water levels, etc.) The study of karst landforms is complex. Special techniques were developed in France more than a century ago and have been used in Belgium, China, Costa Rica, France, Madagascar, Malta, Sarawak, Turkey, and the USA.

Course scheme (Duration 6 hours)

Presentation
Origin and structure of limestone. Action of water.
Karst landforms and karst systems.
Worldwide examples of karstic areas in different climate zones.
The economic importance of tourism.

Tourism
Worldwide examples, economical importance for disinherited areas.
Main problems in caves open to the public.

Geotechnics
The detection of caves in building areas.
Protecting building foundations from the effects of caves.

Natural record-keeping
Genesis of cave formations and natural underground damages.
Dating methods for speleothems and isotopic information.
Natural recording of seismotectonic effects in endo and exokarst.
Natural recording of paleoclimates.

Hydrogeology
Karstic hydrological systems and their functioning.
Study of karstic basins: structural analysis of limestone.
                                               aquifer measure of water flows.
                                               physical-chemical parameters at springs.

different techniques for dye test.
Different examples of water supply in karst and their problems:
                              karstic springs.
                              wells and drilling.
                              underground dams.
                              salt contamination in marine karsts.
Hydrogeological mapping.

5.3 ANNEX III. IMAGE PROCESSING APPLIED TO REMOTE SENSING AND GEOGRAPHICAL INFORMATION SYSTEMS (GIS)

This course provides an introduction to problems linked to the acquisition, processing and management of digital images with particular focus on applications to the Earth Sciences. It comprises both teaching sessions and laboratory work with hands-on initiation to classification techniques and mapping. It is augmented by a series of conferences offered by application specialists and a visit by a leading company in the field.

The aim is to show the synoptic view of the earth seen from space and the challenging problems generated by the enormous quantity of observation data. The course can offer a clear understanding of the many facets of image processing both in terms of physics of measurements, visualisation techniques and content understanding. Classification methods and digital cartography would also be covered. The main course is followed by a one day hands-on session with processing satellite data using a Windows based PC software.   *Duration: 12 to 24 hours*

210

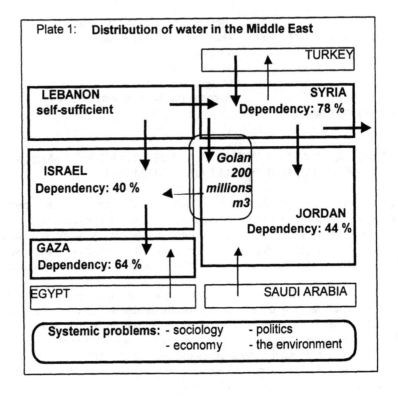

Plate 1: **Distribution of water in the Middle East**

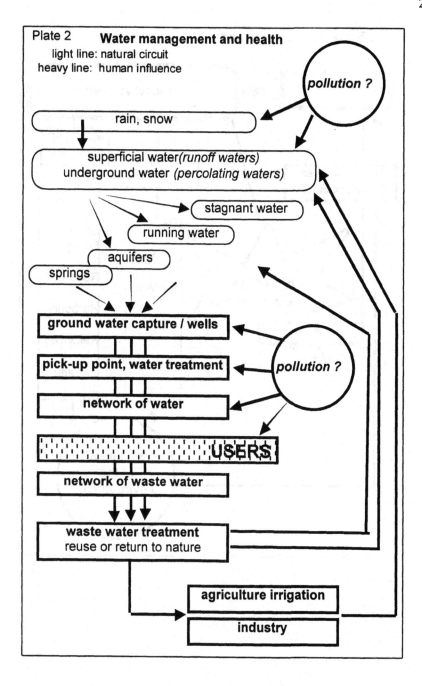

Plate 2   **Water management and health**
 light line: natural circuit
 heavy line: human influence

212

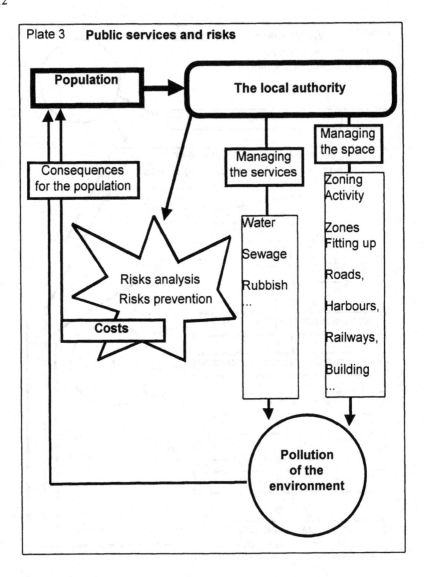

Plate 3 **Public services and risks**

THE STATUS OF TRANSBOUNDARY WATER RESOURCES IN THE BALKANS: ESTABLISHING A CONTEXT FOR HYDRODIPLOMACY

E. VLACHOS
Department of Sociology & Civil Engineering
International School for Water Resources
Colorado State University
Fort Collins CO 80523, USA

Y. MYLOPOULOS
Department of Civil Engineering
Hydraulics Laboratory
Aristotle University of Thessaloniki
54006 Thessaloniki, Greece

1. Introduction

The last half of the 20th century is characterised by significant changes in the planning, design, and management of water resources all over the planet. Mounting concerns about the environmental impacts of human activities, potential climatic shifts, expanding populations and demands, are all expressions of the pressing need to develop alternative institutional schemes for managing scarce natural resources in an integrated manner. The recent United Nations (UN) publication "The Comprehensive Management of the Freshwater Resources of the World" (1997) warns that we must fundamentally change the way we think about and manage water. It also cautions that we must embrace new policies that are comprehensive, participatory, and environmentally sound. Many nations and regions have increasingly been turning their attention to both streamlining existing administrative mechanisms and to introducing innovative institutional arrangements with regard to quantitative and qualitative aspects of their water resources. At the same time, the political significance of water becomes very important, not only because of its scarcity in densely populated regions, but also because of the fact that it is shared across national boundaries.

Up till now, any documented discussion about water contestation has revolved around the central themes of:

J. Ganoulis et al. (eds.), Transboundary Water Resources in the Balkans, 213–223.
© 2000 *Kluwer Academic Publishers.*

- Environmental imbalances and scarcity, result of increasing populations, expanding urban settlements, land degradation, industrialisation, eroded agricultural resource base, etc.
- Hydroculture and a non-co-operative approach, as a result of ethnic, religious, or ideological antagonisms in given river regions.
- Geopolitical settings and power asymmetries among riparian states, especially because of upstream-downstream relationships.
- Particular hydropolitical issues that may intensify particular water conflicts such as dams and diversions.

Thus, there has been quite a clamour for international co-operation and macro-engineering schemes such as in the Danube, Senegal, Mekong, Amazon, Indus, Niger, Nile, Parana/LaPlata and other international river basins. However, as many authors have repeatedly pointed out, international river basins' institutions have largely proven unsatisfactory. Beyond broad legal imperatives and implicit or expressed political will for co-operation, the practice has been predominantly one of no authority, of rare meetings of interested and/or affected parties, of unwillingness to pay, and of little or no information being exchanged.

The issues of sharing international waters and managing the basin resources have thus, become acute in almost every major basin, as well as on many small international rivers previously not subject to any agreement. Examples abound. The Mekong Secretariat has been considering a proposal by Thailand to construct a diversion structure on the Mekong River between that country and Laos for an out of basin diversion. In 1972 India, following a preliminary agreement with Bangladesh, completed the Farraka Barage to divert 40,000 cubic feet per second from the Ganges River into the Bay of Calcutta. Within the last decade, Mexico has complained repeatedly about the increased concentrations of salt in the Colorado River. This has led to the adoption of Minute Orders by the International Boundary and Water Commission (IBMC), in which the United States has developed and implemented a scheme to reduce the level of salt concentrations at the Morales Dam in Mexico. As a result of the 1978 Treaty for Amazonian Co-operation (TCA) member countries formed an association to explore the needs of the countries located on the Amazon River, and to assess environmental concerns. The Nile River, subject of the 1959 treaty between two of the seven riparian states (Egypt and the Sudan), is once more being discussed by other riparian states, which are interested in participating. Slovakia's plan to divert the Danube and build a hydroelectric power station has angered Hungary, which has decried the unilateral changes in boundaries. This situation is leading to an escalation of confrontation between the two nations and even the European Union (EU) is worried that this action may spark off a new crisis in Eastern Europe.

Furthermore, potential climatic shifts, large scale dislocations, socio-economic changes and national and international events have coalesced to create large scale apprehensions and have focused attention on the need for more integrated, anticipatory and far-reaching water policies and strategies. There seems to be an increasing recognition amongst both professionals and ordinary people of the fact that the former relative stability of climate is tending to disappear. Indeed, weather oscillations have increased concern about the present and future impact of long-term climatic changes on the environment.

Experts disagree about how much water is available in given regions. However, there is growing awareness that nations must co-operatively manage, engineer, and conserve available water resources. Indeed, as Gleick [8] points out, no region of the world with shared international water is exempt from water-related controversies, though the most serious problems occur in water-scarce regions. Without co-operative management a zero-sum competition will emerge over water. Seasonal and regional water shortages may exacerbate social tensions and precipitate violence. Sharing and co-operation can provide benefits that exceed those achieved by attempts to maximise individual and national self-interest. Ideally, such co-operation requires a new diplomacy, alternative institutional arrangements, larger financial resources and effective adjudication or conflict management mechanisms.

Recent history, rapid socio-economic changes, socio-political upheavals and the transitions necessitated by the turbulent decades of the 80s and 90s highlight the increasing emphasis given to a variety of environmental challenges. Other issues of increasing importance include the search for sustainable development, the promotion of integrated planning and management, and the attempt to combine structural and non-structural solutions to persistent water resources problems. In this setting of increasing complexity, interdependence and vulnerability, there is an urgent need for intergovernmental integration (through co-ordination, Cupertino and consolidation) of:

- Hydrological interdependencies in terms of both use (rural, urban, industrial, recreational, etc.) and water regime (surface and ground water, quality and quantity).
- Political interdependencies in terms of both horizontal co-ordination in space, and vertical co-operation between levels of government units.
- Transboundary interdependencies, representing both social and hydrological trans-state interdependencies.
- Exogenous interdependencies, most notably the potentially dramatic impacts and consequences of climatic shifts and emerging hydrological alterations.

Many of the water issues that appear at the beginning of the 21st century include items that have been of concern to planners for many years, such as adequate water supply, groundwater quality, and the provision of safe water. Others are of more recent origin, such as acid precipitation, nonpoint source pollution, or adequate instream flows for fish and wildlife protection. What has been definitely recognised, however, is the widely shared perception that water resources problems cannot be solved without taking into account the complex interrelationship of social, political and economic factors.

2. On Hydrodiplomacy

The comprehensiveness of water resource planning has long been the subject of controversy and debate. It has been recognised, however, that in order to maximise the benefits from any water resource project a much large systemic analysis of the surrounding environment is needed. This constitutes a broadening of the traditionally narrow planning and management approaches, and an increased sensitivity to decision-making problems associated with multi-objective and multi-purpose actions. In this vein, a recent publication of the World Bank (1993) elaborated the need for a comprehensive framework for analysing policies and options where water scarcities exist, inefficiencies persist and environmental damages are becoming apparent. In such a context analyses at a river basin level become part of a national strategy for water resource management that includes elements for formulating public policies, regulations, incentives, public investment plans and environmental protection (as well as the interlinkages among them).

The creation of a water management regime may facilitate the emergence of mutually beneficial agreements among those involved on a local or regional level, ensuring that friction does not lead to conflict. An everyday definition of a regime is a Amode or system of rule or government. In contemporary international relations theory, however, regimes consist of procedural and normative guides to state behaviour. Areas in which international regimes have grown most recently include outer space, the oceans, telecommunications and trade commodities. Various types of international regimes have existed for centuries in such areas as the law of the sea, monetary affairs, trade, and communications. However the technological, political and economic forces that have brought nations into these more recent webs of interdependence have extended deeper than ever before into provinces that had once been the exclusive jurisdiction of national governmental policies.

Three international legal organisations of high repute have conducted empirical studies of State practice, on the basis of which they have drafted sets of draft rules for the non-navigational uses of international water resources. The Institut de Droit (Institute of International Law) (IIL) drafted and approved the 1961 Salzburg Resolution on the Use of International Non-Maritime Waters and the 1979 Athens Resolution on the Pollution of Lakes and Rivers. In a similar fashion, the International Law Association (ILA) drafted and approved the 1996 Helsinki Rules on the Uses of International Rivers and the 1982 Montreal Rules on Water Pollution in an International Drainage Basin. Finally, the International Law Commission (ILC), an independent UN legal organisation, was commissioned in 1970 to prepare an authoritative set of rules to be adopted by the UN General Assembly. In July 1994 the ILC completed its draft articles on The Non-Navigational Uses of International Watercourses, and recommended that the articles be elaborated into a convention by the UN General Assembly or an international conference of plenipotentiaries.

Although there are some subtle differences between the IIL, ILA, and ILC drafts, all three contain the same fundamental principles. It is useful to summarise the five major legal principles that are shaping and will further affect hydrodiplomacy practice. These are:

- The principle of international water and the concept of an international watercourse.
- The principle of reasonable and equitable utilisation, a principle that has generated interminable debates and interpretations of the terms reasonableness and equity.
- The obligation not to cause significant harm, and the exercise of due diligence in the utilisation of an international watercourse.
- The principle of notification and negotiations on planned measures.
- A duty to co-operate, including the regular exchange of data.

What all the above imply is that apart from discussing traditional areas of concern in water resources, one must also consider conceptual, methodological, and organisational responses in managing complexity and uncertainty in interdependent natural resources systems. There are many documented examples of lessons learned in the form of similar lists from many countries in terms of comprehensive approaches; co-ordination mechanisms; stakeholder involvement; local responsibility and accountability; sustainability and environmental ethics; social equity; decision support systems and risk management; long-range, macro-engineering emphasis; and, trade-off considerations.

Yet despite all this concern and preoccupation with the fragmentation of authority and the need for thoughtful, far-reaching integration, relatively little systematic analysis has been done as to what institutional mechanisms have been successful or what management schemes appear appropriate in concrete cases. The fragmentations of institutional jurisdictions among different levels of government, between the physical dimensions of water, between various uses and users, and among different sectors have all compounded the difficulties for comprehensive theory and practice. No doubt, integrated management is a popular conceptual framework, but it is difficult to explain as it contains too many concepts, dimensions, and linkages. In whatever form one wants to present integrated management (and some authors refer to it is adaptive or predictive management) it becomes obvious that any pragmatic approach needs to emphasise:

- The recognition of its changing character from crisis management (reactive) to risk (proactive) management.
- The need for interdisciplinary integration that would incorporate multi-objective, multi-action and multi-perspective emphasis.
- The establishment of flexible organisational frameworks, which would allow for increased capability in managing large socio-technical systems.
- The facilitation of the flow of information in terms of monitoring, assessment and evaluation, ultimately allowing for a more efficient, effective and equitable system of management.

- Consideration of the level and scale of intervention, reflecting concern with the degree of centralisation and as to what now appear as central questions of subsidiarity in the EU.

River basin planning and management has long been an honoured tradition, from the development of the Tennessee Valley Authority, to the Senegal and Mekong development plans of more recent years. The underlying, common thread in all such efforts has been the development of water resources for a variety of beneficial uses. From such obvious premises, the fundamental question has been raised as to whether such integrative regional water plans can be fitted within the geographic limits of a whole river basin or watershed. The questions of scale, boundary, and geographic planning unit are of central concern for efforts of definition of the problem, identification of interested parties, and implementation mechanisms. Can such joint planning and management take place in the vast expanses of the Nile, the Amazon and the Parana/LaPlata, or should they be restricted to the more regional, specific socio-political, conflicts of well-defined geographic, cultural, environmental, physiographic or economic boundaries? This is why some authors prefer to talk more about problemsheds rather than watersheds.

The operational terms for international law and other conflict management mechanisms are "complementarity" and "implementation". At the same time, in existing multilateral agencies there seem to be three vexing problems. One has to do with the historical and cultural inertia of past differences and practices. The second with the calculation of all costs involved in the shared waters' development, and the third with the incorporation of social and environmental concerns into planning. These broad problems emphasise the need for negotiation, for third party expertise and for dialogue based on factual information. The agreements that have emerged from efforts in hydrodiplomacy reflect, in great part, the desire of the signatories to engage in comprehensive national planning, basin-wide management, multi-purpose development and water quality control. This constitutes firstly tangible evidence of an increasing concern with the need to craft flexible but durable regimes capable of enhancing the protection of the ecosystem; and secondly a desire to serve the interests of all parties involved by reducing uncertainty, stabilising expectations, and promoting the resolution of conflict as a routine process.

3. Reflections on the Balkans

The press has regularly covered issues relating to the formerly communist Balkan countries in the last four years. Along with former Soviet Central Asia and the Caucasus, it exhibits all the hallmarks of suffering from collapse of the old-world order syndrome. The term Balkanisation has come again to signify political fragmentation, civil wars, bloody confrontations and internecine ethnic fights. The establishment of new states, persistent ethnic enclaves throughout the region, and contested frontiers (sometimes literally between villages), has complicated water independencies and challenged established water management schemes. The partition of Yugoslavia is now at the centre of significant rearrangements, bringing up historical border disputes essentially along the invisible line of the old Austro-Hungarian Empire. To the north and west of this line a central European culture prevails, and to the south and east are areas that existed for centuries under the Ottoman Empire.

Such interdependencies have produced certain conflicts in the last 50 years or so. In the case of the Evros/Maritza/Meric River, which forms the natural border between Bulgaria, Greece and Turkey, there are no major water supply problems as there are no other water uses besides irrigation. However, in the summer of 1993 Turkey strongly protested about diminished water supplies due to hydroelectric power plants in Bulgaria. However, new tensions are expected to rise in the near future, due to low groundwater levels in the region and seawater intrusion in the coastline as a result of high irrigation demands. Of course, the maintenance of the estuary of the river, which has been characterised as one of the most important wetlands and environmental habitats in the world and is protected under the Ramsar convention, remains a big challenge in the area. The only upstream transboundary river in Greece is the Aoos, between Greece and Albania. There have been protests on the part of Albania due to the construction of a large dam on the Greek side. Since 1965 water resources problems in the Axios/Vardar River between Greece and Yugoslavia have increased, due to intense irrigation and accelerating pollution. However, the agreement between Yugoslavia and Greece, which was signed more than 25 years before the partition of Yugoslavia, had quite effectively contributed to maintaining the traditionally good relationship between the two countries. Of course this agreement no longer exists today, as a new country, the Former Yugoslavian Republic of Macedonia (FYROM), is now the second interested party. Meeting the increased water demands downstream, preserving the quality of the Thermaicos Gulf and maintaining the delta of the Axios River, another environmentally sensitive wetland protected by the Ramsar convention, are among the first priorities of a new potential agreement.

Another challenge in the region is the Nestos/Mesta Rivers between Greece and Bulgaria. Despite earlier agreements, Bulgaria has been withholding water supply for its increased agricultural and industrial needs. Since 1975 the Nestos flow has declined from 1,500 million CM to 600 million CM, resulting in repeated Greek protests. A series of negotiations since 1965 have resulted in a new agreement between the two countries, which suffers however from essential weaknesses. According to the agreement, Bulgaria is obliged to leave 28% of the river discharge downstream, but there is no further specification as to the seasonal variation of this amount of water. Another issue not properly treated in the agreement concerns the quality of the water, and more specifically pollution from Bulgarian heavy industry (including thermal pollution from a Bulgarian nuclear plant). Such concerns have raised the level of tension in a region of Greece highly dependent on irrigated agriculture.

There are problems related to water interdependencies in the northern Balkans, especially as they relate to aspects of water quality of tributary rivers to the Danube. A dramatic example is that of the dam built at the Tara River in Montenegro called the a Balkan Chernobyl. This dam holds back a lake of lethal mud containing lead, zinc, mercury, cyanide, arsenic and other poisons from nearby zinc and lead mines. The Tara is one of the fastest flowing rivers that joins the Drina which in turn flows into the Danube, ending up eventually in the Black Sea. This ecological bomb is ticking away behind a crumbling dam. Efforts have been initiated by the UN and the EU to help in shoring up this dam, but it still remains one of the most lethal sources of extensive water pollution in the whole of the Balkan region.

There are also potential problems in the utilisation of three transnational lakes in the region: Prespa (FYROM, Albania, Greece), Ohrid (FYROM, and Albania), and Doiran (FYROM and Greece). Plans for hydroelectrical power dams upstream of Lake Ohrid would be affected if irrigation is undertaken around Lake Prespa or if water from the latter is diverted to Greece for hydroelectric power development. Greece in particular faces not only heavy demands for irrigated water in the Macedonian plains, but also water and power for its cities especially the metropolitan areas of over-urbanised Athens and the conurbation of Thessaloniki. Finally, any scheme to utilise waters from Lake Prespa would involve not only the surrounding countries but also the EU and international environmental groups, since Prespa is an important station for migrating birds. Greece, as the downstream state of four of these river systems, has a particular interest in water supplies, as these comprise about 30% of its freshwater supply.

Thus, a preliminary list of characteristic transboundary water problems in the Balkans includes:

- Poorly defined (and in some cases still in flux) frontiers, especially around former Yugoslavia.
- Significant water quality problems from hazard industrialisation and earlier lack of environmental protection.
- Absence of detailed agreements in many transnational river basins and lack of appropriate administrative/enforcement mechanisms.
- Lack of comparable data and long-range historical records.
- Continuous ethnic rivalries and historical antagonisms with a lingering spirit of mistrust and suspicion.
- Periodic droughts, especially in the southern Balkans, forcing uneven withdrawal of water supplies.
- Significant lack of knowledge as to underground waters, their quality and interdependencies in the region.
- Increasing demands for water due to heavy tourist development e.g. in Dalmatia and coastal Greece.
- Absence of co-ordinating transboundary commissions.

Water, then, as a focus of attention in the Balkans is part of five interrelated and interacting crises:

- A water supply and demand crisis that represents a predominantly engineering dimension.
- Deteriorating water quality crisis that can be translated into an ecological dimension of water problems.
- Transboundary dependencies crisis representing a geopolitical dimension, not only in terms of international frontiers but also intra-national transfers of water across administrative boundaries.

- An organisational crisis exemplified in a <u>management</u> dimension, i.e., appropriate personnel, facilities and procedures as well as legal mandates, court decisions and administrative guidelines.
- A <u>data</u> and information crisis, not only in terms of availability, validity, reliability or comparability, but also as part of combining data and judgement, modelling, and the building of useful Decision Support Systems.

International relations have become so complex that alternative dispute resolution means have become important in the managing or resolving of inter-societal conflicts. The search for alternatives to legal institutions to arbitrate disputes has been prompted not only by a saturation of legal mandates, but also by increasing litigation and confrontation. Mediation, as a compromised discussion between disputants aided by a neutral third party, whose judgement is respected, has become a viable alternative to adversarial processes. The whole range of issues, such as adjudication, arbitration, mediation, conciliation and even principled negotiation, expresses various alternative processes of dispute resolution. However, criticisms have also been made as to whether such processes can compensate for inequitable power relations or can provide incentives for compliance or acknowledgement of the third party decision when there is no recourse to legal sanctions.

As the international scene turns to questions of sustainable development, the restoration and rehabilitation of degraded environments, and the creation of new co-operative arrangements that centre around shared water resources, it becomes apparent that institution building, comprehensive management and alternative dispute resolution efforts will be central quests in the years to come. Diminishing or degraded water resources and their potential impacts on international security, provide unique opportunities for co-operative institutions and for co-operative transnational behaviour. Thus, there is a need for bringing back an environmental approach that requires drastic measures of ecological rehabilitation, innovative institutional mechanisms, and a balance between autonomy and co-operation. Such global approaches also entail improvement in environmental monitoring and information, by expanding the factual basis of comprehensive river basic models. In addition, they also imply a framework for negotiations, which stresses the importance of comprehensive institutional formats and clarity in national and international decision making processes.

Given such considerations and strong socio-political divisions (even centrifugal forces and fragmentation in many nations) there are three responses that should be considered. First improve efforts towards the utilisation of hydrodiplomacy in terms of understanding alternative dispute resolution and conflict management efforts to transboundary water resources. Second, recognise again the river basin approach as a co-operative mechanism and authority, and one that is much more sensitive to ecosystemic interdependencies, and third, place particular emphasis on integrated water resources management (including the building of more robust water resources institutions).

What are, then, the desiderata of an ideal system of managing transboundary river basin conflicts (and for that matter in approaching in a holistic matter water resources development)? Such a system or approach should place particular emphasis on:

- Improving environmental scanning, i.e. expanding the factual basis, the monitoring of trends and the forecasting of social and technical developments.
- Facilitating organisational mobilisation by encouraging innovative administrative units of personnel, facilities and procedures, sensitive and responsive to transnational collaboration.
- Promoting Decision Support Systems (DSS), and the hardware and software that would improve decision-making. Shifting from a current crisis to a risk management style that emphasises both anticipatory and participatory ethos and commitment.

Finally, what needs to be emphasised is an overall strategy of vigilance. This would approach questions of shared water resources and potential contestation flashpoints in terms of: flexible responses, i.e. operational and strategic flexibility; proactive commitment, in terms of environmental scanning and through an emphasis on risk rather than crisis management; river basin focus and robust transnational "regimes"; combinations of global approaches and national plans; ecosystemic emphasis and environmental interdependencies; and, finally, integrated, comprehensive management, capacity building and organisational mobilisation.

4. References

1. Birnie, Patricia W., and A.E. Boyle. 1992. International Law and the Environment. Oxford: Clarendon Press.
2. Carbonneau, Thomas E. 1989. Alternative Dispute Resolution. Urbana: University of Illinois Press.
3. Carroll, John E. 1983. Environmental Diplomacy: An examination and a prospective of Canadian-U.S. transboundary environmental relations. Ann Arbor: The Univ. of Michigan Press.
4. Choucri, Nazl (ed.). 1993. Global Accord: Environmental Challenges and International Responses. Cambridge: The MIT Press.
5. DeVilliers, Marq. 1999. Water Wars. London: Weidenfeld and Nicholson.
6. Frey, Frederick. 1993. "The Political Context of Conflict and Cooperation Over International River Basin" Water International, 18(1):54-68.
7. Gleick, Peter H. 1993. "Water and Conflict: Fresh Water Resources and International Security." International Security 18(1):79-112.
8. Gleick, Peter H. 1993. Water in Crisis: A Guide to the World's Fresh Water Resources. New York: Oxford University Press.
9. Haas, Peter M., et al. (ed.). 1993. Institutions for the Earth: Resources of Effective International Environmental Protection. Cambridge, Mass: The MIT Press.
10. Lonergan, Stephen C. and David B. Brooks. 1994. Watershed: The Role of Fresh Water in the Israeli-Palestinian Conflict. Ottawa: International Development Centre.
11. Mostert, Erik. 1998. "A Framework for Conflict Resolution." Water International. Vol 23, pp. 206-215.
12. Ohlsson, Leif (ed.). 1995. Hydropolitics: Conflicts Over Water as a Development Constraint. London: Zed Books.
13. Rose, Laurence. 1993. "Shared Water Resources and Sovereignty in Europe and the Mediterranean." IBRU Boundary and Security Bulletin, 1(3):62-67.

14. Sands, Philippe (ed.). 1994. Greening International Law. NY: The New Press.
15. Susskind, Lawrence E. 1994. Environmental Diplomacy: Negotiating more effective global agreements. NY: Oxford University Press.
16. Trolldalen, J.M. 1992. International Environmental Conflict Resolution. Washington, DC: NIDR.
17. Wolf, Aaron T. 1997 "International Water Conflict Resolution: Lessons from Comparative Analysis" Water Resources Development vol 13, pp. 333-365.

A PLAN FOR A BALKAN WATER-ENVIRONMENT CENTRE

J. A. VERGNES
Consultant, UNESCO
32 rue Pigeonnier
13300 Salon-de-Provence, France

1. Introduction

The NATO Advanced Research Workshop, which brought scientists and water resources specialists together from all ten Balkan countries, emphasised the need for a Balkan international water project. Such a project would encourage networking among the countries in the future, with the goal of improving and protecting the endangered water resources of the region. This article explores the need for such an approach and suggests ways in which it might be achieved over the long term.

1.1 BALKANS AND WATER

The Balkans are a mountainous region ("Balkan" is the Turkish word for mountain) with a geopolitical definition and history that is both difficult and delicate, and whose borders and number of countries have frequently changed over the centuries. In the broadest sense it is worth noting that the Balkan region has a population of 110 million and that most of its water resources are international. Generally, its water resources are sufficient for current and projected usage. Some areas, such as the Greek islands, have water shortages. In Albania a dry period occurs in the summer when its main watercourses have insufficient flow, while a period of excess water with heavy, torrential flows frequently takes place in the winter. The main regional water problems are caused by pollution.

1.2 BALKANS AND WASTE

In this document, "waste" refers to solid waste, and liquid and gaseous waste emissions. Problems of waste and resulting pollution exist at different stages in every country without exception. Western society has unlimited consumption objectives that generate a great deal of pollution. If there is no change in lifestyle, pollution will endanger the future of life on Earth. Change of lifestyle means a significant reduction in energy consumption per inhabitant in the rich countries and for all new agricultural practices, and a new policy of growth compatible with sustainable development. Urban solid

J. Ganoulis et al. (eds.), Transboundary Water Resources in the Balkans, 225–233.

waste is sometimes partially burned in the street, or often indiscriminately discharged in the environment, on the soil or in rivers. Sometimes, the sewage treatment networks are outdated or not in working order.

Inadequate investment in pollution control is primarily a result of limited national economic development. As in most countries, it also seems that there are no guidelines or sufficient motivation for reflecting on:

- The "wild" garbage dumps that are not based on any minimal sanitary or ecological standard.
- The need for preliminary treatment of products and possible recycling.
- The consequences of a lack of a national environmental policy.

2. The Effect of the War in Kosovo on International Water Resources

War in the Balkans has damaged or destroyed major water networks such as water pipe lines, sewers and watercourses, and associated infrastructures such as monitoring stations, treatment plants and bridges

Even countries not involved in the recent wars suffer from the ecological consequences of pollution and water management difficulties. Economic consequences also exist, in particular from interrupted transport on international watercourses that has resulted in a substantial decrease in the volume of trade between Balkan countries.

In addition to normal pollution, Kosovo has had to deal with the pollution caused by a technological war as well as by the "medieval" practice of well poisoning. Some thirty organisations are currently on record as working to resolve Water and Salubrity (good health) problems. The United Nations High Commissioner for Refugees (UNHCR) is responsible for co-ordinating information about the problems. A study on the water supply issue in all the regions was made for UNHCR in a report made available in August 1999[1]. This study contains information on the human factor in the management of a serious situation, current distribution networks, utilisation of technical treatment, and the maintenance of products reserves such as chlorine, aluminium sulphate, sand filters. Some fifteen non-government organisations (NGOs) are in charge of restoring wells, which are all polluted.

Kosovo's case shows the importance of NGOs' field experience in the framework of the "Water and Salubrity" approach. This expertise should be taken into account in future projects. An example of such expertise is a Greek NGO, HELMEPA, financed by ship-owners and sea manufacturers, which put a lot of effort into cleaning the Greek coastline. The important mission of NGOs in the world is mentioned in Agenda 21 of the Rio Summit (1992).

3. Resolving Water Pollution Problems

Typical water pollution problems include the following:

- Failure of agencies and institutions responsible for water and waste issues to co-ordinate their activities and to communicate with each other.
- Understanding that difficulties in environmental problems are linked to economic, political and cultural issues rather than to the lack of application of technical solutions, including available industrial technologies.
- The urgent need experienced by all countries and the international community to resolve problems of water distribution, quality, and treatment and the drainage, collection, treatment and storage of waste.

Solutions require a common rationale. In 1992, the Rio Congress stressed that environmental issues could be solved only at an international level by using co-ordinated strategies, and not at a national level. The European Union (EU) has integrated a concept of sustainable development, fundamentally important for the future of populations. For these reasons, the EU has invited 15 Member States to take the necessary measures to reduce the degradation of natural resources. The EU particularly encourages environmental audits to better control pollution and its overflow, and provides support to stronger environmental policies among several countries.

It is recommended that:

- A national policy on "Water" cannot be separated from policy on waste treatment.
- A policy connecting "Water and Salubrity" should be based on:

  ❖ A national and regional approach.
  ❖ A multidisciplinary research of solutions.
  ❖ An accountable policy for society using a strategy of information access and sensitisation of the population and the education of children.
  ❖ An evaluation of the national and regional water resources characteristics, of their consumption and pollution (causes and effects).
  ❖ A financing strategy.
  ❖ A policy for training of experts and technician according to needs.

4. Water Production and Distribution in the Balkans

4.1 A PRINCIPLE

A "Water and Salubrity Policy" and not just a "Water Policy" is recommended. One that relates water management to the general well being of society and is not solely concerned with water management. No water production and distribution policy or corresponding study can be separated from a global policy of waste management (urban, industrial and agricultural) or from an environmental protection policy. This principle is important in the Balkans, since although water resources are sufficient in quantity, they are often polluted by the omnipresence of waste and polluted water. As a result water becomes an important carrier of pollution and epidemics.

4.2 SUGGESTED COMPONENTS FOR SOLUTIONS

In order to fully understand each country's particular water issue and to have an overall picture of the water issues in the Balkans with the ultimate goal of initiating a project offering solutions, each country should:

1. Conduct simultaneous research on the existing situation on "Water" and "Waste":

• Prepare a directory of public and private organisations, including name, address, name of responsible person to contact, especially of anyone in charge of training, research, and education centres.
• Provide access to information and training emphasising the need for greater responsibility.
• Determine a general policy on water supply and waste management that will provide:

❖ An evaluation of data on resources (watercourses, ground water sheet, treatment, monitoring units, networks plan) and water consumption (industrial, agricultural, urban, average/inhabitant).
❖ An evaluation of data on waste (causes, effects, collection, final waste) and the state of the treatment and monitoring facilities and distribution networks.
❖ A determination of a viable maintenance, management and administration policy.
❖ General information on budget and financing sources.
❖ Fiscal measures which include incentives for environmental preservation.
❖ Staff training programmes for research/training/research centre.

2. Identify the problems to be solved at urban, industrial and agricultural levels, and in particular emergencies

3. Implement a continuous information and education campaign which repeatedly stresses that water is not an inexhaustible godsend or eternally pure (the tap is a daily miracle), and that safe water resources are already exhausted in some regions of the world, so that:

• By using incentives populations may be encouraged to change their lifestyles and practices.
• A responsible communal policy is developed which prevents pollution, involving those concerned in any decisions leading to changes in habitual practices in agriculture, health, hygiene etc.
• A regional approach is adopted in all studies examining the problems raised.
• A primary-level education project is adopted with the method and strategy adapted for use in each country. (Children are always good bearers of messages to their parents.)

4. Establish what is needed to achieve the above by:

- Training technical and administrative personnel at all levels.
- Training experts especially in law, economic and communication fields.
- Recycling courses and updating knowledge.

5. Propose a general policy project and/or a long-term plan to bring to the fore the needs, priorities, necessary means and financing to allow a rehabilitation and/or amelioration of existing systems.

5. Creation of an International Network of Water-Environment Centres for the Balkans (INWEB)

5.1 THE CULTURAL FOUNDATION OF THE CENTRE

The creation of INWEB fits in with UNESCO Director-General F. Mayor's dictum that "It's time to act and not to discuss". With a view to being useful and efficient, the colloquium should at least propose a project that the participants will promote, using actual communication mechanisms available.

One of the recommendations of the International Water Congress of Kaslik [2] could be accepted as a basic principle for this regional water centre. The Kaslik Congress was held under the auspices of UNESCO with 40 countries and 20 international organisations participating. Its principle can be stated as follows:

- Any teaching or research on water issues should be multidisciplinary. In this context, social and human sciences should form the subject of targeted teaching and research in international law, especially as it applies to international water resources, water economy, education of children, and adult sensitisation. This policy will help to guarantee the success of any environmental project and any agreement between countries.

5.2 JUSTIFICATION FOR THE CENTRE

Justification for the centre lies at several levels. It is needed:

- To promote concerted regional policies. A successful strategy achieved by trainees of different nationalities who initiate dialogue during a training course.
- To answer specific requests and give advice on matters such as training, information, organisation and research.
- To create information sources and an active experts' network in order to help each country's decision-makers.

5.3 CREATION OF THE CENTRE

The creation of the Centre requires a preliminary project that will accomplish the following:

5.3.1. Preliminary study

A preliminary study on the needs of each country should facilitate the creation of this centre. Some regional studies are found on the Internet (EUROWATERNET, for example). Each national study will be communicated to the network of centres via the Internet.

5.3.2. Existing information and reality

In general, countries have incomplete information on their environmental state (water, air, ground, underground and biodiversity) and it is difficult to measure their level of sustainable development. Water data in particular are sometimes old, disputable or even of unknown origin.

Causes of pollution are mentioned but never really measured, for example: the growth in national electric power, urban population, industrial production, tourist population, number of petrol vehicles and waste production.

Information needs to be obtained and stored electronically for easy access. Such information would include legal matters (measures promoting environmental protection, international accords), information on the economy (water taxation, "polluter-payer" measures, "user-payer" measures.), information on the social sciences for cultural and political effects, and on information retrieval, allocation of responsibility among government and non-government organisations, and research and education institutes.

Lastly, studies of countries' environmental policies, their comparison, their evolution, their application and their impact would also be very useful for decision-makers.

5.3.3. Support from a regional follow-up committee

A Regional Workshop Follow-up Committee, created by participants, shall have as its task the creation of an International Water-Environment Centre for the Balkans.

The first job of each Committee member shall be to appoint a national centre (or Institute). This national centre will be the country representative in the network of centres.

Afterwards, each Centre will begin to collect national water numerical data: name and address of water specialists (e.g. research workers interested in water) in exact sciences (e.g. hydrology), social sciences (e.g. law, economics) and human sciences (e.g. sociology, adult education, geography, history).

It will also be necessary to develop a working relationship with other national institutions interested in water issues. This last activity will progressively promote a national multidisciplinary approach, which is important national objective.

5.4   THE OBJECTIVES

When all this national information has been collected, it should support:

- The realistic achievement of the centre's main objectives.
- The identification of common needs in the field of training.
- The initialisation of a broad Balkans database on the subject of water which relates it to the well being of the community.

A number of other topics should be discussed. These include the concept of a "virtual centre" on the Internet with a network of centres; the physical location of INWEB; start up funding, and the choice of expert(s) to define statute, budget, staff, finance sources, strategy, communication devices, training/education curricula etc.
The objectives of INWEB are:

- To make available a directory of all Balkan water centres for training, research and education (address, director, main experts, specialities, publications, etc); to help them develop and to co-ordinate some common activities in order to build a network of centres. Each centre will have a speciality and a mission in the network. INWEB is to be a network of centres, a virtual centre, and a centre with distributed responsibilities.
- To create and manage a database on "Water and Salubrity" and to document the environmental problems of the Balkans.
- To implement the co-ordination of common projects for the Balkans for training technical staff and for other specific requests, and to follow up any such projects.
- To develop the access information system, in order to allow the creation of a database and distance learning, as well as a network of international experts open to the Balkan states.

To summarise, INWEB is a network of centres where each centre will have a mission. This network could progressively ensure the achievement of certain goals: updating of knowledge, training, distance learning, development of the use of the access information system, workshops organisation, water law research, expert network, water data bank for the Balkans, and many other rewarding activities.

6.   Final Report and International Community Help and Participation

A final report should make it possible to seek technical and financial assistance from the international community. Europe will be the prime target as the project concerns the Balkans. This report must show the importance of workshop decisions in the sustainable development of Balkans countries. Recommendations and papers will constitute an important annex to the report.
Europe is involved in these issues. All the Mediterranean countries are dependent on their neighbours as far as pollution is concerned. Moreover, these European countries are united in the implementation of solutions to water issues.

First phase:
This "Final Report of Colloquium of Thessaloniki" will allow those institutions that facilitated organisation of the international meeting to be solicited for technical and financial assistance. According to the colloquium recommendations, such assistance must ensure the creation of an International Water-Environment Centre for the Balkans, and finance for three years.

Second phase:
It would be possible to submit this document to other organisations for additional funding or technical help. These might include:

- The EU and especially the European Task Force Environment/Water (DG12).
- UN organisations such as the FAO, UNEP, UNESCO, UNICEF.
- The World Bank.
- The World Health Organisation.
- The World Meteorological Organisation.
- Other private national and international foundations and funds.

A review of current projects in several countries shows that there should be no major problems in obtaining financial help. For example, the World Bank is providing US$ 100 million for water and the environment in the coastal provinces of North Vietnam. In Vietnam again, UNICEF gave US$3 million for the construction and development of drinking water resources. In Ethiopia, the French Development Agency gave US$ 10 million for the implementation of a drinking water supplying project. France gave US$ 9 million to Kenya for the rehabilitation of hydraulic infrastructures. [3]

7. Conclusions

To sum up the major issues of concern to the workshop and to the Balkan water centre project we should consider the following:

- Technology and data collection alone cannot resolve water issues. Hydrologic investigations cannot resolve cultural or official conflicts, or sensitise a population to the need for improved water management. Education, which will develop a protective water culture, and international legal support are also vital key features of a "modus vivendi" for regional geo-policy. The international water centre will also concern itself with these disciplines.
- Solutions to water issues are dependent on government decisions. The success of any one solution depends on the level of responsibility of the decision-maker. The centre should initiate regional research on these issues.
- Regional multidisciplinary experts and technical staff training projects are crucial. Their efficiency is linked to a good knowledge of the regional context. The centre not

only aims to achieve this, but will also create a network of multidisciplinary experts and a "Water and Salubrity" data bank.

• In the Balkans, as in others regions, the "water crisis" cannot be solved by one action alone and certainly not by a declaration of intent or by research on "*future vision*". But what action should be taken? Not just the implementation of solutions requested by those suffering from a water shortage, but also the promotion of a culture of peace through a water culture.

INWEB will contribute to the implementation of solutions to water issues and promote regional sustainable development. Experts agree that any country's foreign policy should also take water issues into consideration. Both the EU and the UN have integrated the concept of sustainable development into their programmes recommending that:

• Pollution should be avoided at source by using recyclable resources.
• Tax incentives should be available for the private sector, industry and other fields of economic activity preserving the environment.
• The "polluter-pays" principle should be applied to emission control.
• Environmental audits should be used for better pollution control.
• Environmental responsibility should be encouraged: European funds are available to support specified innovative actions.

These organisations should support the above suggestions, whose objectives are to reduce natural resources degradation as well as to develop a culture of peace in this region of the world.

8. References

1. UNCHR, by Reynolds and Grewcock
2. International Congress of Kaslik on Water, Final Report (Jean Vergnes), Kaslik, Lebanon, June 1998
3. Sources: Hydroplus, International Water Review-9/1999 - 33, rue de Villiers -92300 Levallois-Perret - France

Part V: Consensus on a Strategy

CONCLUSIONS AND RECOMMENDATIONS: AN INTERNATIONAL NETWORK OF WATER-ENVIRONMENT CENTRES FOR THE BALKANS (INWEB)

J. GANOULIS
Aristotle University
Thessaloniki, Greece

1. Background

Altogether 50 Balkan specialists on water gathered in Thessaloniki on October 11, 1999 at a hotel high above the hilly port city on the Aegean, to discuss transboundary water problems, and to agree on solutions. They came with expectations; all of them recognised a clear need to restore resources damaged by war and by neglect. This was the first time all ten Balkan countries, including the present Yugoslavian Federation and the four former republics, Greece, Turkey, Albania, Bulgaria and Romania, had gathered for such a purpose. The agenda was full of high priority items from coping with war and environmental damage on the river Sava to protecting the coastal waters of the Aegean from untreated waste. The last agenda item, to find a way to work together for the long-term, was preceded by intense discussion. Tension was relieved, in the traditional Balkan way of scientists and technicians, by humour.

The countries represented at the five-day workshop, which had support from the NATO Scientific Affairs Division and Thessaloniki's Aristotle University, have recently emerged from a decade of war and many years of environmental neglect. Each country first described their current transboundary river, lake, and coast water monitoring systems and provided brief summaries of their condition. Their presentations included maps, charts and tables. Confidentiality was not a factor. They agreed on three items: the shared water resources needed clean-up; priorities for that could not be set until all countries sharing a river basin, lake or sea measured the level of pollution in the same way; and since they were not yet doing that they had better soon find a way to begin and continue such measurements.

2. Workshop Results

The preceding chapters tell the story of the workshop. After each country had reported on the current status of its transboundary waters, western specialists talked about

J. Ganoulis et al. (eds.), Transboundary Water Resources in the Balkans, 235–238.
© 2000 *Kluwer Academic Publishers.*

possible aids to transboundary networks The participants met in four work groups, each one each to discuss solutions to the problems of either rivers, lakes or coastal waters, or the possibility of a Balkan oriented water centre to decide and carry out long-term transboundary water solutions.

The following is a summary of the conclusions that received support from every country.

• National water resources management and environmental protection policies should be based on: a transbounday approach; participation should be multidisciplinary; national water resources should be evaluated in terms of availability and use, with proper attention to the need for a sound financing strategy; training policies should be devised for specialists and technicians with attention to specific transboundary needs, and the legal and social policies of each country should support access to data and information, the prevention of pollution, and the education and sensitisation of citizens about these issues, beginning in the first year of primary school.

2.1 AN INTERNATIONAL NETWORK OF WATER-ENVIRONMENT CENTRES FOR THE BALKANS (INWEB).

The representatives agreed that, in order for recommendations for monitoring transboundary water resources in the Balkans discussed at the meeting to be useful and effective, the ARW should propose a specific project to be promoted by all participants through state-of-the-art information and communication technologies, namely the establishment of INWEB.

The basic principles agreed by the ARW participants for the international/regional water network of centres are the following:

• Any important activity on water issues should be regional, multidisciplinary and aim to protect the environment. It should be in accordance with international laws and treaties, socially and environmentally sensitive, and include reference to the education and sensitisation of children and adults. Protection and rehabilitation of water resources are essential to human health and economic development in the Balkans.

The justification for INWEB was based on a set of high priority needs:

• Existing disparities between national monitoring systems in the Balkans do not provide the scientific and technical basis for an adequate, cost-effective, integrated planning of water resources in the region.
• Existing damage by war and years of neglect of water supply systems is severe.
• Existing weakness of institutional and administrative structures in a multitude of small international water basins, which were created after the collapse of the former Yugoslavia, persist and interfere with solutions.
• Existing or potential conflicts on the use of water in the region need resolution.
• There is a need to create a reliable and objective information source in the Balkans and an active experts' network in order to support government policy effectiveness.

- Globalisation of the economy and new communication and information technologies have created the need for new ways of working based on networking, collaboration, and the ability to anticipate future changes.
- Funding from the European Union (EU) and other international organisations may be more easily granted to a regional co-operative group rather than to individual institutions.

2.2 INWEB OBJECTIVES

The objectives of INWEB are to:

- Facilitate the exchange of information in the fields of water and environment by establishing an international, open network in the Balkan region.
- Contribute to public education in the fields of water and environmental protection.
- Promote training and professional development in the Balkan region by providing distance learning.
- Develop an inventory of existing transboundary monitoring systems.
- Create and maintain a database on Water and the Environment by use of official, validated national data complementary to the EUROWATERNET, related to transboundary water bodies (rivers, lakes, and coastal waters).
- Establish an international expert network open to the Balkan States.

2.3 APPROACH

The following should be taken into consideration:

- INWEB should have a management committee with one representative from each country with a rotating Chair, and an executive secretariat.
- INWEB needs funding from national and international organisations, including the EU.
- A detailed feasibility study should precede a formal proposal for the IWCB.
- The legal status of INWEB will be defined in the feasibility study
- Some preliminary investigations of information needs in each country should be part of the study. Such national studies should be based on existing national data and institutional structures avoiding duplication of other initiatives.
- From these studies and other regional studies available on the Internet proposals will be made on the following subjects: technical terms of reference, geographical boundaries, national representation, sources of funding and the legal constitution of INWEB.

- The participants at the workshop agreed to establish a Regional Follow-up Committee, whose mission will be to help develop the feasibility study and to participate in INWEB.

3. Conclusions

In the Balkans, as in other regions, there will be no solution to water crises unless action is taken. The basis for action is as follows:

INWEB, as proposed by this workshop, will contribute to the improved management of the region's water resources. It will develop common projects to train technical staffs, provide access to information and data and sustain a network of experts. The EU, the United Nations (UN) and other international organisations have accepted the concept of sustainable development. It is requested that these organisations support these recommendations of natural resources as well as the development of a "Culture of Peace" in the Balkan region.

LIST OF PARTICIPANTS

NATO ADVANCED RESEARCH WORKSHOP
TRANSBOUNDARY WATER RESOURCES MANAGEMENT IN THE BALKANS: INITIATING REGIONAL MONITORING NETWORKS
11th-15th October 1999, Thessaloniki, Greece

(*) Member, INWEB Follow-up Committee

ALBANIA

Prof. Dr. Skender Osmani
Faculty of Geology and Mining
Rr. Elbasanit
Tirana
Tel: 355 42 752 45/46
Fax: 355 42 423 72
e-mail sosmani@yahoo.com

Ass. Prof. Dr. Veli Puka
Hydrometeorological Institute
Surresi Street 219
Tirana
Tel: 355 42 – 23518
Fax: 355 42 – 23518
e-mail velipuka@hotmail.com

Prof. Dr. Mitat Sanxhaku (*)
Hydrometeorological Institute
Durresi Street 219
Tirana
Tel: 355 42 23518
Fax: 355 42 23518
e-mail: mitats@yahoo.com

Dr. Lirim Selfo
National Environmental Agency
Lake Ohrid Conservation Project
Tel: 355 42 42738
Fax: 355 42 42738
e-mail: lselfo@icc.al.eu.org

BOSNIA AND HERZEGOVINA

Prof. Tarik Kupusovic (*)
Hydro-Engineering Institute
Stjepana Tomica 1
71000 Sarajevo
Tel: 387 71 207949
Fax: 387 71 207949
e-mail ihgf@utic.net.ba

BULGARIA

Assoc. Prof. Roumen Arsov (*)
University of Architecture, Civil
Engineering. and Geodesy,
Faculty of Hydrotechnics
1, Chr. Smirnenski Blvd.
1421 Sofia
Tel: 359 2 66 995
Fax: 359 2 668 995
e-mail: r_arsov_fhe@uacg.acad.bg

Assistant Prof. Ilka Lambova
Faculty of Mathematics & Informatics
Sofia University
5 James Bourchier Blvd.
1164 Sofia
Tel: 359 2 6256584, 6256565
Fax: 359 2 687180
e-mail: ilambova@fmi.uni.sofia.bg

Assoc. Prof. Mariana Maradjieva
University of Architecture, Civil
Engineering and Geodesy,
Faculty of Hydrotechnics
1, Chr. Smirnenski Blvd.
1421 Sofia
Tel: 359 2 656 863
Fax: 359 2 656 809, 656 863
e-mail: marmar_fhe@uacg.acad.bg

Assoc. Prof. Dr. Georgi Mirinchev,
Ministry of Environment and Waters
NCESD
136 Tzar Boris III blvd.
1618 Sofia
Tel: 359 2 955 9818

Prof. Dr. Valentin Nenov
Bourgas University
Bourgas
Tel: 359 56 660021
Fax: 359 56 686141
e-mail: vnenov@btu.bg

CROATIA

Prof. Jure Margeta
Faculty of Civil Engineering
University of Split
Matice hrvatske 15
21000 Split
Tel: 385 21 303 356
Fax: 385 21 524 162
e-mail: margeta@cigla.gradst.hr

Prof. Ognjen Bonacci (*)
Faculty of Civil Engineering
University of Split
Matice hrvatske 15
21000 Split
Tel: 385 21 303 340
Fax: 385 21 524 162
e-mail: obonacci@gradst.hr

FRANCE

Prof. Roger Casanova
University of Nice-Sophia Antpolis
Surville, F.83111 Ampus
Tel: 33 4 94 70 97 28
Fax: 33 4 94 70 97 35
e-mail: casanova@unice.fr

Prof. Jean Vergnes
UNESCO Consultant
32 rue Pigeonnier
13300 Salon-de-Provence
Tel: 33 4 90 56 54 54
e-mail: Jean.Vergnes@wanadoo.fr

F.Y.R. OF MACEDONIA

Oliver Avramoski
Lake Ohrid Conservation Project
96 000 Ohrid
Tel: 389 96 263 743
Fax: 389 96 263 743
e-mail: oliverav@iunona.pmf.ukim.edu.mk

Dr. Zdravko Krstanovski (*)
Hydrobiological Institute
96 000 Ohrid
Tel: 389 96 262 910
Fax: 389 96 262 810

Prof. Dr. Igor Nedelkovski
Faculty of Technical Sciences
St. Kliment Ohriodski University
I.L. Ribar bb,
97000 Bitola
Tel: 389 97 31 355
Fax: 389 97 48 320
e-mail: nigor@osi.net.mk

Zoran Spirkovski
Hydrobiological Institute
Lake Ohrid Conservarion Project
96 000 Ohrid
Tel: 389 96 262 810
e-mail: zoranspi@rsc.com.mk

GREECE

Prof. J. Ganoulis (*)
Department of Civil Engineering
Hydraulics Laboratory
Aristotle University of Thessaloniki
54006 Thessaloniki
Tel: 30 31 99 56 82
Fax: 30 31 99 56 81
e-mail: iganouli@civil.auth.gr

Prof. Yannis Krestenitis
Department of Civil Engineering
Hydraulics Laboratory
Aristotle University of Thessaloniki
54006 Thessaloniki
Tel: 30 31 99 56 5654
Fax: 30 31 99 5658
e-mail: ynkrest@civil.auth.gr

Assoc. Prof. Yiannis Mylopoulos
Deptof Civil Eng. Hydraulics Lab.
Aristotle University of Thessaloniki
54006 Thessaloniki
Tel: 30 31 99 5695
Fax: 30 31 99 5658
e-mail: mylop@civil.auth.gr

Athanasios Pantrakis
Periferia
An. Makedonias &Thrakis
Tel: 30 531 21625,31319
Fax: 30 531 25989

Epaminondas Sidiropoulos
Dept. of Rural & Surveying Eng.
Aristotle University of Thessaloniki
54006 Thessaloniki
Tel: 30 31 99 6143
e-mail: nontas@topo.auth.gr

Ioannis Stympiris
Periferia
An. Makedonias &Thrakis
Tel: 30 531 22 861
Fax: 30 531 25 989

As. Prof. Marios Vafiadis
Department of Civil Engineering
Hydraulics Laboratory
Aristotle University of Thessaloniki
54006 Thessaloniki
Tel: 30 31 99 5685
Fax: 30 31 99 5658
e-mail: vmargaritis@civil.auth.gr

Prof. El. Papachristou
Department of Civil Engineering
Hydraulics Laboratory
Aristotle University of Thessaloniki
54006 Thessaloniki
Tel: 30 31 99 56 44
Fax: 30 31 99 56 81

ROMANIA

Prof. Radu Damian
Tech University of Civil Eng.Bucharest
Bd. Lacul Tei 124
Sector 2
72302 Bucharest
Tel: 40 1 242 1163, 242 1208 /ext. 163
Fax: 40 1 242 0781, 242 0777
e-mail: damian@hidro.utcb.ro

Graziella Jula
ICIM,
Spl. Iudependentei 294,
Sector 6
77703 Bucharest
Tel: 40 1 221 5758
e-mail: gjula@pcnet.pcnet.ro

Enfg Mihaela Lazarescu
Researc & Eng. Inst for Environment
Independentei 294
Sector 6,
Bucharest 78
Tel: 40 1 221 09 90
Fax: 40 1 221 9204
e-mail: icim@nfp-ro.eionet.eu.int

Dr. Petru Serban (*)
Water Management, Hydrology and
Meteorology Division Apele Romane
Str. Edgar Quinet nr. 6 Sector 1
70106 Bucharest
Tel: 40 1 3122174
Fax: 40 1 3122174
e-mail: serban@ape.rowater.ro

RUSSIA

Prof. Vladimir Evstigneev
Altai State Technical University
46 Lenin Ave.
Barnaul 656099
Tel: 7 3852 26 14 34, 26 09 17
Fax: 7 3852 26 05 04
e-mail: vve@api.altpi.altau.su

Svetlana Fetissova
Altai State Technical University
46 Lenin Ave.
Barnaul 656099
Tel: 7 3852 36 78 43

Dr.Irina Mikhailidi
Altai State Technical University
46 Lenin Ave.
Barnaul 656099
Tel: 7 3852 35 48 94
e-mail: ingem@rocketmail.com

Yulia Samosudova
Altai State Technical University
46 Lenin Ave.
Barnaul 656099

Denis Smoline
Altai State Technical University
46 Lenin Ave., Barnaul 656099
e-mail: sdv@agtu.secna.ru

Oleg Solodki
Altai State Technical University
46 Lenin Ave., Barnaul 656099
e-mail: solo@agtu.secna.ru

Prof. Alexander Tskhai
Altai State Technical University
46 Lenin Ave., Barnaul 656099
Tel: 7 385 2 36 7038
Fax: 7 385 2 36 7038
e-mail: taa@agtu.altai.su

SLOVENIA

Prof. Mitja Brilly (*)
University of Ljubljana
Faculty of Civil and Geodetic
Engineering, Hajdrihova 28
1000 Ljubljana
Tel: 38 6 61 125 3324
Fax: 38 6 61 219 897
e-mail: mbrilly@fagg.uni-lj.si

Dr. Jasna Grbovic
Hydrometeorological Institute of
Slovenia, Vojkova 1b
1000 Ljubljana
Tel: 38 6 61 327 461
Fax: 38 6 61 133 1396
e-mail: jasna.grbovic@rzs-hm.si

TURKEY

Prof. Atil Bulu (*)
Istanbul Technical University
Faculty of Civil Engineering
80626 Maslak
Istanbul
Tel: 90 212 285 3735
Fax: 90 212 285 3710
e-mail: bulu@itu.edu.tr

Prof. Dr. Derya Maktav
Istanbul Technical University
Faculty of Civil Engineering
80626 Maslak, Istanbul
Tel: 90 212 2853808
Fax: 90 212 573027
e-mail: maktaav@srv.ins.itu.edu.tr

Prof. Dr. Ing. Ahmet Samsunlu
Istanbul Technical University
Faculty of Civil Engineering
80626 Maslak
Istanbul
Tel: 90 212 2853784
Fax: 90 212 285 6587 / 2853781
e-mail: samsunlu@itu.edu.tr

Assoc. Prof. Delia Teresa Sponza
Dokuz Eylul University
Engineering Faculty
Environmental Eng. Dept.
Buca, Kaynaklar Campus
Izmir
Tel: 90 232 453 1153
Fax: 90 232 453 1153
e-mail: dsponza@izmir.eng.deu.edu.tr

Assoc. Prof. Aysegul Tanik
Istanbul Technical University
Faculty of Civil Engineering
Environmental Eng. Dept.
80626 Maslak, Istanbul
Tel: 90 212 285 6884
Fax: 90 212 285 3781
e-mail: tanika@itu.edu.tr

U.K.

Tim Lack
Water Research Centre
Henley Road
Medmenham, Marlow,
Bucks SL7 2HD
Tel: 44 1491 636 590
Fax: 44 1491 579 094
e-mail: lack@wrcplc.co.uk

UKRAINE

Vladimir Kuznetsov
Ukrainian Scientific Research Institute
of Ecological Problems
6 Bakulina Str
310166 Kharkiv
Tel: 380 572 45 50 47
Fax: 380 572 45 50 47
e-mail: mnts@uscpw.kharkov.ua

U.S.A

Prof. Darrell Fontane
Colorado State University
Department of Civil Eng. Fort Collins,
CO 80525
Tel: 1 970 491 5247
Fax: 1 970 491 6787
e-mail: fontane@engr.colostate.edu

Dr. Irene L. Murphy
2005 37th Str., N.W., Washington D.C.
20007 USA
Tel: 1 202 337 2376
Fax: 1 202 342 6434
e-mail: imurph@aol.com

YUGOSLAVIA

Prof. Dejan Ljubisavljevic (*)
Faculty of Civil Engineering
University of Belgrade
B. Revolucije 73
11000 Belgrade, Yugoslavia
Tel: 38 1 11 321 8557
Fax: 38 1 337 0223
e-mail: ljubisav@irc.grf.bg.ac.yu

Prof. Slobodan Vukcevic
Institute for Technical Research
Cetinjski Put B.B.
81000 Podgorica, Yugoslavia
Tel: 38 181 214 456
Fax: 38 181 213 961

GLOSSARY

AUT	Aristotle University of Thessaloniki
B&H	Bosnia and Herzegovina
BAS	Bulgarian Academy of Science
BOD	Biochemical Oxygen Demand
BRI	Biological Research Institute
BSS	Bulgarian State Standard
CIP	Civil Engineering Faculty
COD	Chemical Oxygen Demand
CWME	Croatian Water Management Enterprise
DAEWS	Danube Accident Emergency Warning System
DCP	Data Collection Platforms
DGXI	Directorate General XI, European Commission
DO	Dissolved Oxygen
DPSIR	Driving forces, Pressures, State of environment, Impact on environment, and Responses in the form of policy and regulations
DRC	Department of Reservoirs and Cascades
DSI	State Water Works
DSS	Decision Support Systems
ECTS	European Credit Transfer System
EEA	European Environment Agency
EIEI	Administration of Electricity Surveys
EIONET	European Environmental Information and Observation Network
EPA	Environmental Protection Act
EPA	U.S. Environmental Protection Agency
EPDRB	Environmental Programme for the Danube River Basin
ESA	European Space Agency
ETC/IW	EEA and its Topic Centre on Inland Waters
EU	European Union
FR	Federal Republic
FYROM	Former Yugoslav Republic of Macedonia
GAEP	Southern Anatolia Environmental Project
GEF	Global Environment Facility
GIS	Geographical Information System

GTS	Global Telecommunications System
HBN	Hydrologic Benchmark Network
HBI	Hydrobiological Institute
HMI	Hydrometeorological Institute
HPP	Hydroelectric Power Plants
HU	Hacettepe University
IBMC	International Boundary and Water Commission
IEA	Integrated Environmental Assessment
IIL	Institute of International Law
ILA	International Law Association
ILC	International Law Commission
INWEB	International Network of Water-Environment Centres of the Balkans
IPH	Institute of Public Health
ITU	Istanbul Technical University
IWM	Integrated Water Management
MAF	Ministry of Agriculture and Forestry
MAFWE	Ministries of Agriculture, Forestry and Water Economy
MAP	Mediterranean Action Plan
ME	Ministry of the Economy
MED-HYCOS	Mediterranean Hydrological Cycle Observing System
METU	Middle East Technical University
MEV	The Ministry of the Environment, Physical Planning and Public Works
MEW	Ministry of Environment and Water
MLIM	Monitoring, Laboratory and Information Management
MH	Ministry of Health
MRDPW	Ministry for Regional Development and Public Works
MUPCE	Ministry of Urban Planning, Construction and Environment
MWFEP	Ministry of Waters, Forests and Environmental Protection
NASQAN	National Stream Quality Accounting Network
NAWQA	National Water Quality Assessment Programme
NCESD	National Centre for Environment and Sustainable Development
NEA	National Environment Agency
NEAP	National Environmental Action Plan
NGOs	Non-Government Organisations
NIMH	National Institute of Meteorology and Hydrology
NOAA	National Oceanic and Atmospheric Administration
NRL	National Reference Laboratory
NWS	National Weather Service
PE	Population Equivalents
PHI	Public Health Institute
PIAC	Principal International Alarm Centre

PIAC	Principal International Centre
PMU	Project Management Unit
PPC	Public Electric Power Corporation
RAS	Russian Academy of Sciences
RHMI	Republic Hydro-meteorological Institute in Skopje
SS	Suspended Solids
TAC	Treaty for Amazonian Co-operation
TNMN	Transnational Monitoring Network
TUBITAK	Turkish National Scientific and Technical Research Centre
TWPCR	Water Pollution Control Regulation
UN	United Nations
UNHCR	United Nations High Commissioner for Refugees
USGS	U.S. Geological Survey
WFD	Water Framework Directive
WHO	World Health Organisation
WMMP	Water Management Master Plan
WMO	World Meteorological Organisation
WPMP	Water Protection Master Plan
WQRP	Water Quality Ranking Plans
WWF	World's Wildlife Foundation
YTU	Yildiz Technical University

WEBSITE CONTACTS

Global

UNDP   http://www.undp.org

UNEP   http://www.unep.org

UNESCO   http://www.unesco.org

Europe

Mediterranean   http://www.unep.org/water/regseas/medu.htm

Black Sea Environmental Program http://www.dominet.com.tr/blacksea

Danube Programme   http://www.rec.org/danubepcu/

Danube hydrology   http://www.datanet.hu/hydroinfo/vituki/info/index.htm

Rhine Commission   http://www.iksr.org

Altai Region   http://www.ab.ru

European Environment Agency   http://www.eea.eu.int

European Topic Centre on Inland Waters   http://etc-iw.eionet.eu.int

USA

Historical water quantity/quality   http://water.usgs.gov

Real-time, maps   http://water.usgs.gov/streamgaging

Water data   http://waterdata.usgs.gov/nwis-w/US/

Water information and data   http://waterweb.org

Weather, climate   http://www.noaa.gov/

Environmental quality   http://www.epa.gov/enviro

Bureau of Reclamation   http://www.usbr.gov/power/data/data.htm

Balkan Countries

FYR of Macedonia

Faculty of Natural Sciences and Mathematics
Sts. Cyril and Methodius University, Skopje

http://www.pmf.ukim.edu.mk/

Greece

AUTH   http://www.auth.gr/

INWEB   http://socrates.civil.auth.gr/inweb/

Ministry of Environment,
Public Works and Planning   http://www.minenv.gr/

Turkey

MP/TR   http://www.ins.itu.edu.tr/imp/medpro.htm

Slovenia

Ministry of environment and spatial planing legislation
http://www.sigov.si/cgi-bin/wpl/mop/vsebina/angl/index.htm

Hydrometeorological institute data
http://www.rzs-hm.si/
The Chair for hydrology and hydraulic engineering with connections
http://ksh.fgg.uni-lj.si/ksh/

INDEX